You turn things upside down,
As if the potter were thought to be like the clay!
Shall what is formed say to him who formed it,
'He did not make me'?
Can the pot say of the potter,
'He knows nothing'?

(Isaiah 29:16, NIV)

Biblical Creation
and the
Theory of Evolution

D. C. SPANNER

Exeter
The Paternoster Press

AUSTRALIA:
Bookhouse Australia Ltd.,
P.O. Box 115,
Flemington Markets, NSW 2129

SOUTH AFRICA:
Oxford University Press,
P.O. Box 1141,
Cape Town

Quotations from the Revised Standard Version © 1946,
1952, 1971, 1973 by the Division of Christian Education of
The National Council of the Churches of Christ in the
United States of America.

The picture (cover, page 18) of the Creation
is taken from a Latin Bible of 1511, printed in Venice,
and is reproduced from an original in Bible Society's
library at Cambridge University Library,
by permission of the Syndics.

British Library Cataloguing in Publication Data

Spanner, D. C.
  Biblical creation and the theory of
  evolution.
  1. Creation 2. Evolution——Religious
  aspects——Christianity
  I. Title
  231 7'65   BS651

  ISBN 0–85364–315–6

Typeset by Photoprint, 9–11 Alexandra Lane, Torquay
and Printed in Great Britain for The Paternoster Press,
Paternoster House, 3 Mount Radford Crescent, Exeter, Devon
by A. Wheaton & Co. Ltd., Exeter, Devon.

TO GWENLLIAN
~ my dear wife ~
not always scientific
but always a true helper

Bible Translations

JB   Jerusalem Bible
NEB   New English Bible
NIV   New International Version
RSV   Revised Standard Version
RV   Revised Version

Contents

Preface

This little book is the result of many years' study and thought on the issues with which it deals. I have been a lifelong reader of the Bible, and have sought to allow it to mould my thinking and my behaving. Since graduation I have also been a professional scientist, my chair in the University of London having been of Plant Biophysics. Five years before I retired in 1978 I was ordained to the non-stipendiary ministry of the Church of England. I therefore have a foot in both camps, and can (I hope) view from a reasonable angle both the defensive obscurantism of many upholders of the Bible's authority and the blinkered over-confidence of many of those whose authority is (they suppose) established science. My own convictions are of the God-givenness of Holy Scripture and of the God-givenness of Science (as a discipline), each in its sphere an avenue to understanding. The problem is their right use, and it is as a contribution to establishing that use that this book has been written. It may be helpful to describe briefly its plan. The first and larger section is devoted to setting out what (I believe) the Bible teaches on the questions at issue. This has to be done fairly thoroughly because ignorance and misunderstanding, even among the educated, are here so widespread and profound. In the course of this exposition some challenging and unorthodox conclusions are reached. There follows a discussion of some technical points relating to evolutionary theory, points which are very significant in the present context but which are usually dismissed with scant attention. Then there is a brief statement outlining current neo-Darwinian orthodoxy—brief because the subject receives comparatively good publicity to-day. Finally there is an attempt to meet the obvious challenge: how can the biblical doctrine and the findings of science be related harmoniously together? A number of specialized topics, important in themselves, have been relegated to Appendices to avoid cluttering up the main argument.

It remains only for me to say how much I owe to the constructive thoughts of others, mostly members of the Research Scientists' Christian Fellowship. If I had to single out one name it would be that of the late Professor Donald MacKay, to whose faith, integrity, insight and lucidity I owe more than I can say. I thank God on every

9

remembrance of him and of all the men and women of faith I have known among my scientific colleagues. In other directions I should like to thank Mrs Ann Summers, the friend who typed my manuscript; and my son Huw who saved me from a number of grammatical and logical infelicities. It only remains to commend this modest effort to Him who alone can make it of use in the extension of His kingdom.

DOUGLAS C. SPANNER

CHAPTER ONE

Introduction

The greatest need of our contemporary world is to recover the conviction that God is Creator. How do we know that he is? Because of the biblical testimony. Why should we accept this? An answer is given addressed both to scientifically-minded readers and to fellow-clergy. The presuppositions of the argument are set out.

When an author sits down to write a serious book he needs to be clear in his own mind about three questions: What is his purpose? To whom is he addressing himself? Is the subject worth the effort? To these he might add a fourth: What are the pre-suppositions of his argument? Let me therefore, as much for my own benefit as for my reader's, say something at once about these matters.

My purpose can be stated very simply. It is to commend the biblical doctrine of God the Creator. Is the subject important? Certainly. In fact, I would go so far as to say that a renewed grasp of this truth is the most urgent need of modern sophisticated man. We live our lives (at least in the West) at a desperately superficial level. Few of us stop to ask: "Has existence any objective, built-in purpose or meaning?" Yet men and women recognize deep, even profound, inner needs and cravings, and search restlessly and endlessly for ways to satisfy them. The outcome is usually that their energies are channelled, for the most part, into seeking pleasure and what is euphemistically called a 'higher standard of living'. Alas, their failure to find what will satisfy is self-evident. Our society is continually eroded by discontent. Prices and wages spiral upwards year by year. Meanwhile the problems we face become ever more grave: the disparity between rich and poor, the drug traffic, the pressure from over-population, the increase in the power of weaponry, the breakdown of authority, the devaluation of human life, and so on. How shall we face the future? The overriding impression is that most of us prefer not to. We let the fierce unrelenting momentum of life hurry us on without a pause. Hardly a split second separates one item on the 'box' from the next. If people were to stop and think they might realize that they had nothing to live by remotely resembling an intelligent philosophy of

11

life. Instead, they have a routine—with occasional variations. They are journeying aimlessly through existence to a totally unknown destination. In every sense of the word, this is a hopeless situation. Nothing can change it for good so radically and positively as the conviction that life has been *given* to us, and for a purpose: in other words, that God is Creator.

On the face of it there would seem to be every reason for people to wish to subscribe to this belief. But an important obstacle to its reinstatement is a certain unwillingness to be intellectually committed. Most people, it appears, have a native tendency to believe in God. Yet for various reasons scarcely acknowledged, many, especially the 'educated', prefer to remain rather vague in their beliefs, to keep their formulation flexible or plastic. When problems for thought arise their creed can then be reshaped as necessary to avoid confrontation. Thus they escape the discomfort of such stresses as more solid convictions would impose on them. So far so good, we may think. But in losing the problems they lose also the strength that comes from having convictions (to change the metaphor) with teeth and claws. What avoids fighting can never defeat foes, nor win battles. To be worthwhile, therefore, my purpose in writing must not be to commend a doctrine of God as Creator which is nebulous and ill-defined. In fact, it is not. It is to commend the *biblical* doctrine.

This book is addressed mainly to two classes of people: to readers with a scientific outlook and to fellow clergy.[1] Scientists are no more irreligious than other people. But the nature of their work concentrates their attention on questions of *mechanism,* and this very concentration tends to put out of focus questions of *meaning.* Concepts such as 'chance' and 'randomness', for example, have a currency much wider than their narrow technical one. Scientists may easily fall into the trap of thinking about them uncritically, and concluding that a definition which is adequate for doing science is adequate for comprehending life. This can be quite disastrous, and it is for this reason that I have included two or three rather specialist chapters on these topics. In addition, scientists, in common with most educated people today, have a wholly insufficient understanding of what the Bible teaches. It is a profound book, and its doctrine cannot be evaluated at a sitting. So I have tried in the earlier chapters to take advantage of what generations of scholars have concluded about its teaching, and to pass it on.

I have a different reason for addressing my fellow clergy. Today many of them are suffering not only from a crisis of identity but also from a crisis of belief. Ultimately the Christian faith stands or falls with the authority of its foundation documents; of the Old Testament books as the lively 'Oracles of God', and of the whole Bible as 'God's word written'.[2] If the authoritative witness of the Bible is rejected what reason have we after all for believing such an otherwise incredible doctrine as the Incarnation? In this respect a great many clergy seem to have lost their nerve. Whereas G. K. Chesterton (I think) once described himself as 'prancing with belief', they might be described as hamstrung through scepticism. Misunderstanding of Science is partly to blame, and especially misunderstanding of Darwinism. About this I shall have something positive to say later. Destructive criticism of the Bible is even more the cause of their diffidence; and here, not being a scholar, I can say less. But I owe it to my readers to say something at least of why I hold with complete conviction the historic Christian understanding that, in Paul's words, 'all scripture is inspired by God'.[3]

There are of course the standard arguments drawn from the testimony of Scripture to itself, and from our Lord's attitude to the Old Testament.[4] These are of supreme importance, as is the argument from the Bible's own self-evident authority, and the immense power it has wielded, and continues to wield, over thoughtful men and women who read it. But I would like to add three other considerations not so often brought forward.

During the last 150 years, a great deal of scholarly effort has been devoted to the question of how the books of the Bible came into existence. The first five books, spoken of collectively as the Pentateuch and of obvious significance in the present context, have received their full share of attention. The Documentary Hypothesis, still widely canvassed, is built around the suggestion that they result from the interweaving of several sources, usually referred to as J, E, D and P. This theory has raised two issues quite distinct but often confused. The situation it has created is in fact analogous to that resulting, in another sphere, from Darwin's theory.

The first issue is whether the hypothesis is true. This is very far from being settled. It is disputed by scholars of the first rank.[5] So also is the extremely tendentious effort being made to base on it an evolutionistic, reductionistic account of the origin of Israel's religion. This is no more likely to succeed than the attempt to base ethics on Darwin's theory. The base is entirely inadequate for the edifice to be erected on it.

The second issue is whether a naturalistic explanation of origins, however successful, can ever negate the claim that the Pentateuch on the one hand, or the living world on the other, is divinely 'given'; in other words, whether such an explanation can rule out the view that the Pentateuch is the 'Oracles of God', or that the living world is the handiwork of God. Now the claim that a successful naturalistic explanation necessarily excludes God-givenness is based on a fallacy

widely held by both believers and secularists alike. The Bible flatly
denies this claim. For instance, it provides itself, in effect, a naturalistic explanation of the death of Jesus Christ—priestly jealousy, an
individual's resentment, social unrest, foreign domination. Given
these (and a few other common circumstances) the secular historian
would be quite satisfied that he knew why events took the course they
did. Yet the biblical writers robustly affirm that there is a truth taking
precedence over any such naturalistic explanation—that this happening was both in broad outline and in finer detail God's doing, the
central act of his plan for 'reconciling the world to himself'. 'This
Jesus', said Peter, '[was] delivered up according to the definite plan
and foreknowledge of God'. 'Herod and Pontius Pilate conspired
with the Gentiles and peoples of Israel to do', unknowingly, 'all the
things which under thy hand and by they decree, were foreordained'.[6]
   On this analogy therefore it is quite illogical[7] to demand that
genuinely divine revelation must come in the sort of supernatural, i.e.
physically miraculous, fashion in which the Book of Mormon is said
to have been given;[8] or that man's origin must have been likewise
palpably supernatural and miraculous before we can accept that he is
really 'created in the image of God'. The Bible clearly does not
support this demand. It is invalid because it arises out of a false idea
of God; he is much 'bigger' than it allows. He holds, the Bible says,
not only the miraculous but also the ordinary, firmly in his grasp and
disposition. That is why the demand is misconceived. This point is a
most important one. It will crop up again later.[9]
   My next point is rather different. It concerns the nature and
standing of the New Testament message—that 'Jesus Christ . . . the
Lord of Glory', 'died for our sins in accordance with the Scriptures,
that he was buried, that he was raised on the third day', that he is
coming again.[10] Now most men of sense and goodwill (at least in the
West) would agree that if true, this message is of supreme
importance. Indeed, it is hard to conceive of a message which *could*
be of greater importance for a race that is fast losing all sense of
meaning, purpose and destiny. Yet according to not a few leading
scholars and theologians this message has been allowed to reach us in
a form permanent and accessible to public examination, scrutiny and
verification, only through writings which are in vital respects quite
misleading. If their views are right (and I have in mind, for example,
Bultmann's treatment of the Resurrection), then God has apparently
taken very little care to ensure that we ordinary men and women have
an accurate account of his saving actions and provisions. He is surely
therefore open to accusations of indifference or folly or both. I imagine
myself explaining to honest, intelligent but totally unchurched
scientific colleagues the Incarnation, Atoning Death, Resurrection

and Parousia[11] of Jesus Christ. 'This is Good News', I tell them, 'from the God of truth. He has intervened historically in dynamic wisdom, love and power to rescue men. And he's anxious that all men should know'. 'Where do you get your information from?' they enquire. 'From a collection of old writings', I reply, 'the New Testament'. 'Are they reliable' they ask. '—on the historical level, where they can be checked?'. 'Well, no', I reply, 'unfortunately they are in fact of dubious value as historical documents; they contain a good deal about miracles, a virgin birth and a bodiy resurrection which the writers evidently believed but which scholars now discount. There's a good deal of spurious matter woven in almost inextricably too. However, with the help of modern scholarly techniques and existentialist insights the main message is clear enough'. By this time my scientific audience are courteously but frankly incredulous.[12] And no wonder. To them it just doesn't add up. This isn't the way they write-up their important results—arbitrarily interweaving the genuine and the spurious—and they wouldn't expect God to have had his written up like it either.[13] The effort by scholars like Bultmann[14] to make things credible has, in fact, made them incredible.

Now what I have written about the New Testament documents applies to Scripture as a whole, and not least to the Pentateuch. It deals with matters too important, far too closely related to the gospel, to be left to the speculative and uncertain 'insights' of men, however gifted and religiously expert. Experts can differ, and in a crisis like ours nothing could be worse. This brings me to my third consideration. The crisis is upon us. We cannot afford to wait (and, in fact, we never have been able to) for scholar and theologian to achieve certainty before making up our minds on the authority of Scripture. This has never been more obviously true than today. It is arguable that the ethical dilemmas facing us grow continuously larger and more urgent, and the consequences of a wrong choice worse and more irredeemable. We live, as never before, in one world. We need a sure guide, publicly accessible and recognizably authoritative. But God has never been indifferent to our need. He has always taken the common man seriously, and it is entirely in character with his revelation of himself in Jesus Christ (whom 'the common people heard gladly') to believe that he has provided for ordinary people in concert with their fellows and no doubt with help from specialists, a guide[15] at once plain and trustworthy.[16] The great historic confessions of the churches have acknowledged the Bible to be just such a guide. Indeed, that was how our Lord regarded it. In doing so, he quite often used language closely resembling the very formula, 'the Bible says', which periodically (and sometimes derisively) comes under criticism.[17] It is for this reason (among others) that notwithstanding the loss of nerve in the church today I have based my arguments unashamedly on the Bible's teaching. The blasts of theological scepticism, after all, have a way of blowing themselves out, like other storms. Wittgenstein is reported to have said that 'philosophical analysis, if properly done, leaves everything as it is'.[18] I have the feeling that the same will prove to be true of that sort of destructive criticism which makes of the Bible not the Word of God but only the insights (however brilliant) of men.

All that remains now is to set out a little more clearly the presuppositions of my argument. In order to do this, I shall use an old analogy which goes back at least to Francis Bacon. Nature and Scripture are two of the books out of which God instructs men. A great book can be read in different ways. Thus *Oliver Twist* can be read as entertainment, as social comment or as political propaganda. Today the most influential way of 'reading' the Book of Nature is called Science. The corresponding way of reading the Bible is called Theology. My presupposition is that the analogy we are discussing is a valid and valuable one. Let me briefly make some salient points. For Science, its book is *of unquestioned authority;*[19] all scientific doctrine stands under its judgement.[20] The *unity and internal consistency* of this book are taken by scientists for granted, however paradoxical and anomalous its deliveries may seem.[21] Its 'statements' (i.e. the data gathered by scientific observation) are understood in their *prima facie* sense (i.e. 'literally') wherever possible, but sometimes they have to be interpreted in a more sophisticated way;[22] there are no *a priori* rules about this. The book which Science reads is *public property*, accessible to all alike who will take the trouble to qualify themselves to read it. The prime requisite for success, the basic essential, is honesty.[23] Granted this, the understanding of the Book of Nature becomes a co-operative enterprise, in which progress is made through the contributions of different minds sharing a common loyalty to its authority. To minds of earlier ages the book proved comprehensible at the level of men's most fundamental needs;[24] to those of later times it discloses, without end, truths to enrich material life.[25]

My presuppositions should now be plain. In this brief account of Science let the reader replace 'the Book of Nature' with 'the Bible', and 'Science' with 'Theology', (making whatever few grammatical changes are necessary for good sense) and they will be clear enough. In addition, however, this analogy will serve as this scientist's apology for the historic Christian view of the Bible, for when we examine current objections against the historic view of the Bible's authority we find few of them that could not be raised, with similar force, against the scientific approach to Nature. In both cases, the proof of the pudding is in the eating.

CHAPTER TWO

The God of the Bible

What is God the Creator like? Among others, his attributes are personhood, holiness and righteousness, all relevant to the subject of creation. Again, as Spirit, he is not confined within our four-dimensional space-time continuum, since this is itself part of the created order. God's 'abode' must be thought of as something dimensionally much richer. With this biblical understanding of the greatness, majesty and eternity of God the question of the 'six days' of the divine activity in creation assumes quite a different aspect.

'IN THE BEGINNING, GOD . . .'. With these majestic words the Bible introduces us to a truth (as I believe it to be) of prodigious significance: the Ultimate Reality, behind all lesser realities, is personal. For that is what these opening words mean. Man is not 'alone in the unfeeling immensity of the Universe',[1] the product and plaything of titanic physical forces which neither know nor care what they are doing. His existence instead is set before a Presence, august, tremendous, all-seeing.[2] This is the teaching we have to consider.

Clearly, before we can think about the Bible's doctrine of Creation intelligently, we must know something of its doctrine of God, unfortunately no longer a matter of common knowledge. What is God the Creator like? Is it possible to know anything about him? That we have definite but scarcely-acknowledged images in the backs of our minds may well be brought to our attention by looking at the woodcuts in Figure 1 (p. 18). How many, unreflectingly perhaps, enter-tain such pictures as these as illustrating fairly not merely 'mediæval' conceptions of the Creation, but *biblical* ones? Of course they do nothing of the sort. Even the magnificent paintings of Michael Angelo might be seriously misleading, so far as the biblical conception is concerned, unless we are on our guard. It was not for nothing that Moses warned Israel that they 'saw no form' on the day when they received the law at Sinai, lest they attempt to make an image of God.[3] Any such representation would inevitably, he implied, be a misrepresentation. We need to clear our minds at the outset, therefore, of any ideas which may come into this category.

What are the essential characteristics of God the Creator as the

Fig. 1   God's creation of the world in six days, as described in the Book of Genesis. From a Latin Bible of 1511, printed in Venice.

Bible portrays him? Of course he is thoroughly *personal*.[4] 'God said', 'God saw', 'God blessed', 'God called' are some of the recurring themes of Genesis which bring this home. 'Let those who suffer according to God's will . . . entrust their souls to a faithful creator'[5] is

how the New Testament emphasizes it. Then God is *Spirit*. He is
concealed from human observation, however close the scrutiny. Like
the wind he is known by the movements he initiates but he himself
remains unseen and unseeable.[6] Thus he is never to be thought of as
one physically-detectable Agent (even the very greatest) among
many, for Spirit is a different and transcendent order of being.[7] He is
not localized physically in space, as our Lord implies when he tells the
Samaritan woman that God is to be worshipped 'neither on this
mountain nor in Jerusalem', but 'in spirit and in truth'.[8] Again, God
is not localized in time. 'Lord, thou hast been our dwelling place in all
generations . . . from everlasting to everlasting thou art God'.[9] God
fills all space and all time, yet not *merely* that; he transcends them
both. This is the strong impression that is conveyed by such passages
as 2 Chron. 6:18 ('Behold, heaven and the highest heaven cannot
contain thee'); Isaiah 44:24 ('I am the Lord, who made all things,
who stretched out the heavens alone, who spread out the earth'); and
Isaiah 57:15 ('. . . the high and lofty One, who inhabits eternity,
whose name is Holy'). Biblical thought is concrete, not abstract like
that to which Science has accustomed us;[10] in Scripture time and
space are not distinguished from the events and objects that occupy
them. Thus 'stretching out the heavens . . and spreading out the
earth' can very naturally be taken as indicating the establishing of
space as a characteristic of the created order; and the 'heavens . . .
wear out like a garment. Thou changest them' as indicating the same
for time.[11] Similarly the opening verse of Psalm 90, already quoted,
with its mention of God, 'our dwelling place in all generations',
suggests the idea that time and space, the arena in which successive
human generations live out their brief parts, themselves exist within
the mind of God, rather than God and his mind within them. This
was certainly how the great Augustine regarded the Bible's teaching,
as we shall see later.

Two attributes of God often emphasized in the Bible are his
holiness and his righteousness.[12] They are not the same. The word
'holy' in the Hebrew Scriptures probably comes from a root meaning
'separate', 'distinct'.[13] Most often this idea is used in an ethical sense,
but it is also used in connection with creation. When it is so used it
indicates the never-to-be forgotten distinction between God and all
that he has made, between Creator and creature.[14] This is a common
theme in the Bible, and is, of course, an explicit denial of pantheism
in all its varieties. 'Righteousness' is an idea whose main thrust is
again ethical. In man it means conformity with God's moral and
spiritual law which, designed for his good, is the standard or plumb
line by which his conduct is judged.[15] As applied to God himself the
meaning is a little different. It is best expressed as self-consistency;

for God is bound by no standard external to himself. But he is never arbitrary; he abides by his declared policies, and stands by his covenant.[16] Thus one aspect of his righteousness is faithfulness (i.e. abiding by promises he has given or hopes he has raised). Now all this has profound implications for the *physical* aspect of the creation. Righteousness in an important sense implies law, and faithfulness predictability; and it would be strange indeed if these characteristics of the moral order did not apply also to the physical, since God is the Author of both. In fact, the Bible often pointedly links them:

> When he established the force of the wind and measured out the waters, when he made a decree for the rain and a path for the thunderstorm, then . . . he said to man, 'The fear of the Lord— that is wisdom, and to shun evil is understanding.'[17]

It is in keeping with this that the Bible consistently sees the regularities of Nature as dependent not on the perfection of her mechanism, but on the faithfulness of her Creator.[18] Thus it is the same attribute of *righteousness* that on the physical level undergirds the validity of the scientific enterprise,[19] and on the spiritual level fills the horizon with hope.[20]

Other attributes of the biblical Creator—his wisdom, love and power—are more generally recognized and we need not at this point spend time over them. Rather, because of its early relevance to our enquiry, we must return to the idea of God as Spirit, especially in connection with the notions of space and time. This will mean a fairly lengthy digression for which I must beg my reader's indulgence.

The most obvious sense of our Lord's words (in John 4:24) that God is Spirit would seem to be (as noted before) that God is not materially localized. For reasons already given this can be expressed in present-day terms by saying that he transcends space and time. We must probe into this idea a little more deeply. (If the reader finds this difficult he may prefer to go on to the next chapter and come back to this later). We begin with a discussion of space.

Everyone knows that the space we live in is three-dimensional ('3D'). Why *three* dimensional? Why not four, five or even more? The question is not a logically absurd one. Mathematicians regularly use spaces of a large number of dimensions to establish theorems of practical use. But the fact remains that however useful and legitimate such a conceptual extension may be we remain quite unable to form a mental picture of any dimensions beyond our familiar three. The reason for this is that our experience takes us no further. However, some simple considerations may serve to convince us that dimensions beyond these three almost certainly exist.

Consider the simple question: what would happen if we were to travel on indefinitely in a straight line? One answer is that we should go on for ever; the universe is infinite. Men have held this view, but it is a most problematic one.[21] It is doubtful if its import can really be grasped. For the secularist, at any rate, it reduces the sum total of all that is human to total insignificance, in fact to zero. For the theist it has other objections hardly less serious. But the next answer, that we should come to a stop, an edge to space, is no less difficult. What could such an eventuality possibly mean? Relativity theory has provided a third answer, that we should arrive back at our starting point. Space, it suggests, is 'finite but unbounded', curved on the analogy of the two-dimensional surface of a sphere. Now this clearly overcomes the objections to the two previous answers, but it introduces a special requirement of its own; there must be a hitherto unrecognized dimension *into which* our 3D space can be curved.[22] If this suggestion of Relativity be true therefore, our familiar three dimensions of space are not all there are. There is at least one dimension more. Of course, it is impossible for us to visualize it.

The idea of time is more difficult than that of space. Time has, in fact, always been a very thorny subject, even for professional philosphers, many of whom have held it to be unreal. 'The importance of Time is rather practical than theoretical' wrote Bertrand Russell,[23] and that is, in fact, one reason why it is important in the present study, for the Bible is an essentially practical book. Time is always, in some way, concerned with movement. Either it is a stream bearing us along,[24] or it is a track along which we are travelling,[25] or it is something else—moving.[26] We can't so readily objectify it as we can space. It's too intimately associated with our consciousness. We would all agree that it takes time to think; not all would agree that it takes space to do so, in a comparable sense. Nevertheless space and time have many analogous features, and in Relativity Theory they are, of course, regarded as inseparable, forming the well-known four-dimensional space-time continuum.

Time itself is usually thought of as one-dimensional. There is, however, no logical necessity for it to be confined to one dimension. Many of my readers will remember that C. S. Lewis in his famous children's story *The Lion, the Witch and the Wardrobe* used the idea that time may be (at least) two-dimensional. On the return from her first journey into Narnia Lucy apologizes for having been away so long—'hours and hours', she says. The others don't understand; they saw her in the room only a moment ago! Thinking her mad, they tell the Professor, 'Lucy had no time to have gone anywhere, even if there was such a place'. 'That's the very thing that makes her story so likely to be true', said the Professor. 'If she had got into another world I should not be at all surprised to find that the other world had a separate time of its own; so that however long you stayed there it would never take up any of *our* time.' (Something similar, of course, might have been said of its space).

If our discussion so far has done nothing else, I hope it has shown that there is no self-contradiction in believing that there exist dimensions of reality beyond those with which our experience acquaints us. That we are quite unable to visualize them is neither here nor there. Physics has limbered us up to take that hurdle long ago. But what bearing, it may be asked, has this conclusion on the credibility of the Genesis creation story? I hope the answer will become apparent as we go on.

Centuries ago Augustine of Hippo made some profound observations on the subject of Creation.[27] He imagines a 'fickle-minded' man wondering why God 'should have allowed countless ages to elapse' before he undertook the work of creation. What was God doing 'then', before he undertook to create? For if the creation is derivative and not co-eternal with the Creator, then it would seem reasonable to ask what the Creator was doing 'then', i.e. before he called the creation into existence. Augustine's answer is brief and to the point: there was no 'then', for *time itself is part of the created order.* God 'made all time; He is before all time; and the 'time', if such we may call it, when there was no time was not time at all.'

Augustine deals in the same way with space.[28] 'How did You make heaven and earth?' he muses. 'Clearly it was not in heaven or on earth that You made them . . . Nor was it in the universe that You made the universe, because until the universe was made there was no place where it could be made.' So space too is part of the created order. So also, of course, is matter: 'Nor did You have in Your hand any matter from which You could make heaven and earth, for where could you have obtained matter which you had not yet created . . .? Does anything exist by any other cause than that You exist?' Augustine therefore firmly takes the view that space, time and matter are all alike elements of the creation, all alike given their existence by God the Creator.

This view calls for two comments. The more important is that it is thoroughly biblical, as we have seen. It accepts that God the Creator is not confined to our space-time continuum; he operates from his 'holy and glorious habitation', from a realm that is ineffably greater and more wonderful.[29] That he is often spoken of in space-time language ('enthroned in the heavens', 'thy years have no end')[30] in no way alters this conclusion. Even the ineffable has sometimes to be put into words.

The second comment is less important but still interesting. Augustine's view is in close harmony with modern Relativity Theory. To the physicist of today space and time are no longer (as they were for Newton) components of an otherwise empty continuum[31] into which we can think of material bodies with their associated energy being introduced (or from which, withdrawn). They are *essentially* bound up with matter and energy. According to the view of General Relativity, if matter and energy ceased to exist space and time would cease to exist too. Like Augustine, modern theorists regard them as an integral part of the material order.

How then do we relate the biblical Creator to the space and time in which we ourselves exist? No doubt we must regard him as essentially unconstrained by them, as transcending them, however he may choose to act within them. Perhaps the simplest way is to think of him as dwelling in a space and time dimensionally richer than ours, and 'engulfing' ours as three-dimensional space engulfs two-dimensional.[32] This may raise as many problems as it solves, but at least if we adopt it we shall not be limiting God in our thoughts quite so unworthily as we might do. It may help us to see a way through problems which would otherwise be somewhat opaque to us. Its significance, in connection with the topic of Creation, will be developed a little further in a later chapter; but it is by no means limited to this topic, as a familiar incident will show.

John's gospel[33] records an event in our Lord's controversy with the Pharisees. It took place geographically within the temple precincts, and historically at one of the great Jewish feasts, perhaps in the winter of AD 29. It certainly belonged to our familiar space-time. But in connection with it, and almost in the same breath, our Lord refers to an event of a different order altogether: he was 'consecrated and sent into the world'. The natural understanding of this is to regard the 'consecration' as having taken place in God's 'holy and glorious habitation';[34] and the 'sending into the world' as being along an axis not *in* our space-time but *into* it, from outside.[35] The placing of the two types of event in such close juxtaposition here (as in many other places in Scripture) does seem to imply an analogy between them; and granting this, it would seem the analogy can be usefully worked out in terms of dimensionality, higher and lower.

The purpose of this discussion on space and time should now be clear. It has been to help forward the conviction that we have to visualize the theatre of God's activity as Creator not as coincident with our narrow space-time continuum, but as something grander by far. God creates (for creation is continuing, as we shall see) not *in* our continuum, but *into* it. As with Redemption, it is a movement from outside[36] into the arena of space and time in which man lives out his mortal life. With this perspective the problem of the 'six days' in which 'the Lord made heaven and earth' looks quite different. For the way is open, as Augustine long ago suggested, to see the days not as *periods of physical time at all,* but as something far more wonderful. 'Of what fashion those days were, it is either exceeding hard, or altogether impossible to think, much more to speak'.[37] It is possible to see them as days of eternity, not of time: of the life of God, not of the life of man.[38] When one thinks that the subject is nothing less than *creatio ex nihilo*[39] is not this conclusion more or less inevitable? Hasn't it about it an inescapable logic? For, to the biblical writers, God is not one cause (even the greatest) among many. He is the One 'giving to all, life and breath and everything', the 'Source, Guide and Goal of all that is'.[40] They strain the resources of language to express the majesty, the glory, the incomparability of the Creator. Our trouble is that 'our God is too small';[41] to understand the Bible's teaching we need to let God be God. If we can only bring ourselves to do this, as I hope to show, one of the great stumbling-blocks for the modern reader—that of the 'six days' of Genesis—will be removed. Of course, there still remains the question of why the Creator's activity is spoken of in precisely the terms it is, i.e. of 'days' and of 'evening and morning'. That is a point which will be taken up later.

CHAPTER THREE

Approach to Genesis

The Bible is to be thought of as God preaching to his creatures, explaining himself to them and them to themselves. Genesis, therefore, is to be approached as a living communication from God. It is not myth, having a fundamentally different origin and function. Its relation to the creation myths of antiquity is discussed, and the relevance of scholarly theories about its origin to our understanding of its authority and message.

Before we go any further we must pause and ask an important question, or rather a series of questions. They are suggested by the reflection that the interpretation offered here for the days of Genesis 1 is a rather sophisticated one. Did the author, whoever he may have been, really intend the 'six days' to be understood in such a veiled, recondite way? Isn't it wholly anachronistic to introduce the lofty abstractions of up-to-date dimension theory into a primitive account of how the world came into being? Doesn't it seriously misjudge the *genre* to which Genesis belongs, mistaking primitive myth for developed metaphysics, legend for history and poetry for prose? In fact, isn't it altogether too clever? These are legitimate questions, and within the limitations imposed by my lack of biblical and literary scholarship I must do my best to answer them.

The first question to be faced is this: what comprehensive view are we to take of Genesis itself, as a whole? Is it history, or myth, or saga or a mélange of pieces originating separately but skilfully assembled by some literary master? The answer to this will surely be similar to that which we give to the larger question: 'What view are we to take of the Bible as a whole? What is *its* genre, literary and theological?' Many educated people alas, never bother to give themselves a considered reply. They muddle on in their public hearing or private reading hoping that they will somehow come to some suitable conviction. This is most unsatisfactory. It recalls the position of those educated—*theologically* educated—men of old who asked by our Lord, 'John's bapitsm—was it from heaven, or from men?' declined to answer.[1] 'The authority of the Bible—is it from God, or from men?' is, after all, a rather similar question. Can it be that declining to answer it springs from a motive as little worthy as theirs was? The price they paid was to lose the truth. Can anything be worse than that? The attitude of many contemporary scholars is not always so clear as it might be. They earnestly debate whether the first three words of Genesis are to be understood absolutely or relatively; whether the Hebrew preposition of Genesis 1:27 should be translated 'in' or 'as'; whether the 'image' is the same as the 'likeness'; and so on. Do these things *really matter*? Not, surely, if Genesis has only the authority of human genius! If that is all there is to it, learned and prolonged debate on these trifles smacks of mere theological dilettantism. But if Genesis has *divine* authority the case stands quite differently. We ought positively to be eager to know the true meaning, and to regard no labour too great to find it. That is what, in fact, many of our best scholars do; one could wish they would state more clearly why they do it. Do they really regard the Bible as *Revelation*? as a Word before which we must bow? One would like to know. For myself, I have no hesitation in answering 'Yes', to these questions. In the light of our Lord's attitude to the Old Testament,[2] the testimony of Scripture to itself, and the immense splendour and power of these writings (so ancient and yet so contemporary) the only conclusion of permanent value is this: the Bible is God speaking— God explaining himself to men, and men to themselves. It is God preaching[3] to his creatures, and taking as his text now history, now nature, now experience. The Bible informs us that its own great purpose is to show humankind how to live according to the will of God, and Scripture is God himself speaking to us (2 Timothy 3:16; Deuteronomy 29:29b).

Before we go any further there is a fundamental position to be established. Concerning any historical event God has a point of view as well as man; that should be self-evident. Obviously, the two points of view, although *complementary*, are not of equal standing. Again, that hardly needs arguing. God sees things as they really are, i.e. absolutely. Man sees them only as they appear, relatively. Further, man's 'seeing' is often faulty and always tentative. But even when it is clear-eyed, it by no means follows that it will be easy to reconcile it with the divine viewpoint. Where we are dealing with things which are complementary to each other, that difficulty only exemplifies a common state of affairs sometimes experienced with great poignancy, as the physicists know only too well.[4] Now God offers to share his viewpoint with man; that is, in fact, what is meant by revelation. What happens when the revelation seems to fit very awkwardly with man's own view of things? One consequence may be its outright, even indignant, rejection. Our Lord often faced such a reaction, and it is a frequent complaint in the Bible as a whole.[5] A more sensible reaction, in view of our fallibility in matters directly accessible to us and our finiteness in face of things which aren't, is humility and teachableness. These are rare virtues, but ones (so the Bible declares) which God values exceedingly highly.[6] In what way does all this apply to the question we started with: How should we approach Genesis? Let me illustrate by a profound parallel.

The event we call Calvary is, to the professional historian, the entirely understandable outcome of certain historical, sociological, cultic and individualistic factors. A motley assemblage of men and movements were somehow, by the accidents of history, fused together into an uneasy unity: the ambitious chief priests; the Sadducees, scribes and Pharisees; the Herodians, Zealots and Roman soldiery; Pontius Pilate and Herod. An idealistic and non-violent young prophet fell foul of their combined forces. The result was almost predictable, and certainly quite understandable in entirely naturalistic terms. I personally would not wish to dispute this in the least. Yet the Bible tells us that from the point of view of the sovereign Lord all was, in unfathomable mystery, the outworking of a single, co-ordinated divine plan. Further, it tells us that *salvation for men consists of seeing it like this*.[7] To think that the key to the meaning of Calvary lies in the hand of the professional historian—or any other professional for that matter—*is to miss the real significance altogether*—an incalculable loss![8]

Now what is true of Calvary as the Doing of God is true also (I would argue) of Scripture as the Oracles of God.[9] The secular historian investigates the one, the critical scholar the other. Their studies issue in naturalistic reports. Within their own terms of reference these may be both wholly accurate and entirely valid—but they miss the great essential. That isn't their fault: they are tuned (necessarily) to another wavelength. What is inexcusable however (and we are thinking now of Scripture) is when they deny (on the ground of their findings) that God is speaking here and now to humble hearts, in authoritative fashion, in the language of perspicuity, and through the very Bible he has providentially put into our hands, about the great themes of Creation and Redemption, Eden and Gethsemane. That is a denial whose substance we cannot for a moment concede. It takes the Bible as divine revelation effectively out of the hands of the common man. It distances God from men and women and denies him any real significance as a public communicator, one whose very *words* can be shared and pondered together (contrast the emphasis in Isaiah 1:2, Psalms 94:9 and 115:5, and Acts 17:11).

So Genesis is to be approached as the living Word of God. Our next problem is interpretation. How do we interpret it? Much has been written about this,[10] and a number of particular problems are dealt with in other chapters of the present work. Scripture is the great interpreter of Scripture, and with this in mind a point of rather fundamental importance may be raised at once. It is suggested by a sermon which our Lord preached to the crowds at Jersualem when the chief priests challenged him about his authority to cleanse the temple. It is usually referred to as the Parable of the Wicked Husbandmen.[11] We notice two things. It is in the language of metaphor, but its reference is distinctly historical. Our Lord undoubtedly meant his hearers to understand it as setting out the divine perspective on the nation's history, and they took is as such. Israel is the vineyard, its leaders are the tenants, the prophets are the servants sent to receive the fruits and the Messiah—himself—is the beloved Son. In other words, the characters are real historical people. The parable (unlike say the parable of the Sower or of the Prodigal Son) moves in the dimension of actual history. It is not merely existential. It is historically particular. (In passing, note that it is because it is expressed in *metaphor* that it is possible to ask quite wrong-headed questions, such as, 'What system of irrigation was used in the vineyard, and what cultivars?' This characteristic of metaphor needs to be carefully noted, or we shall be easily sidetracked). Now the parable illustrates a mode of speaking very often found in Scripture, and there are many reasons for believing that the narrative of the Garden of Eden, of the two trees and of the final expulsion follows this mode. But (as our example shows) that is entirely consistent with the story setting forth from God's standpoint, real history.[12]

We return to our previous subject. Sermons can be of different sorts, as every preacher knows. Can we be more specific, in particular about Genesis? Are the early chapters 'myth', for instance? Here we need to avoid arguing at cross purposes. The word is decidely ambiguous. To the layman 'myth' means a story without foundation in fact, a mere figment of the imagination, unrelated to truth. To the scholar, however, myth is related to truth, but not to *scientific* truth. The story it presents is not one which could have been video-recorded live, in other words, had the necessary skills been available. It did not happen in history. Its truth is rather existential, or belonging to some higher category still. Now with this understanding of myth what can we say of the early chapters of Genesis? For reasons given elsewhere[13] I regard the story of Eden and the Fall not fundamentally as myth[14] but as history. The account of the creation, especially in the first chapter, is not so easy to categorize. A video-recorder could, one imagines, have captured a great deal of what went on—the movement of the seas and the appearing of the dry land; the sprouting of the vegetation and the swarming of the waters; the increasing of cattle and creeping things; and the coming of man. But it wouldn't have recorded any evidence that it was *God* who was doing these things. How could it? What would the evidence have sounded or looked like? The idea that the tape would have picked up a voice saying 'Let there be light', or a video signal showing the Deity (in some physical shape or other) moulding clay into a human form and then breathing upon it—this idea is too naïve and out of harmony with the biblical conception of God to be entertained for a moment. True, God does appear later in the Bible story in what is technically referred to as 'theophany'; but the situations are very different. His power is always veiled and muted as he condescends to men.[15] So we conclude that the creation narrative does record physically-real happenings, but it also bears witness to the divine authorship above and beyond them. The former are historical; the latter some scholars would call 'mythical'. But the word is too tainted to be accepted without protest. A better word is 'metaphysical'. How we distinguish between the historical and the metaphysical (as between the literal and the metaphorical) is a matter of experience and judgement. There are no *a priori* rules. As we saw in the Introduction, there are similar problems in Science.

Looking at Scripture as 'God preaching' yields positive insights. Let me instance one or two. Every preacher faces a mixed audience; and every good preacher aims to be comprehensible, in his really important matters, to all who are listening. God is the preacher *par excellence*. Thus there is always a meaning in Scripture addressed to the immediate audience, and relevant to the situation in which the message was first given. The prophecy in Isaiah 7 about the young woman who was to conceive was addressed in the first place to King Ahaz, and was intelligible to him where he stood.[16] But the New Testament makes it plain that when God preached that sermon he had more distant horizons in view as well, and congregations not yet born.[17] It had a significance for them that Ahaz did not, and probably could not know; the time was not yet ripe. This is, in fact, something very frequently noticed in the Bible.[18] It is quite reasonable therefore to maintain that the casting of the creation narrative into the pattern of the six days was a device designed both to speak to those contemporary with the writer, and also to be intelligible, in terms more sublte as well, to generations far later. To early Israel (and indeed to men of all ages) the purpose was to establish a weekly pattern of life—six days of disciplined work followed by one of rest and reflection on their covenant relationship with God. To this end it was natural (and intended) that popular Israel should understand the 'days' as the only days they knew, of twenty-four hours. They were right to do so. This understanding taught them how to live according to the will of God, and that is all that ultimately matters.[19] But when God had given mankind further insights into the nature of the physical world, and further training in abstract thought, it became permissible and indeed right to think in more sophisticated terms, and that is the justification for the present attempt. What is so surprising is that this ancient record, understood in its own terms and in the light of the whole biblical revelation, can be set alongside the best of modern scientific thought and still be seen to overshadow it, as something altogether bigger and more all-encompassing. The scientific world-view indeed is seen to fit, as I believe it always will be seen to fit, *inside* the biblical. It is swallowed-up by it! Can anything remotely comparable be said of the great creation myths of the ancient world— the Babylonian or Egyptian for instance?

The mention of these very ancient creation myths raises some further interesting questions. It has been customary in some quarters to regard these myths as primary and the Genesis narrative as secondary, that is, as based on them but improved immeasurably by the genius of Israel's religious thinkers. Certainly the two have interesting elements in common. But it is no more necessary to regard the Genesis account as derived from (say) the Babylonian than it is to regard the Ten Commandments as derived from the earlier code of Hammurapi. C. S. Lewis has a remark[20] that great moralists are sent not to teach people the moral law but to *remind* them of it, 'to restate the primeval moral platitudes'; for men have already an inborn sense of what is right and wrong, defective though this may be.[21] In a rather similar way it is entirely reasonable (on biblical premises) to believe that very early, man had already an inherited understanding of the origin of things, though a sadly distorted one; an understanding to which these ancient myths bear witness.[22] The Mosaic revelation recalled Israel (and the world) to a true understanding. This would account both for the common elements in all these early narratives (including Genesis) and for the enduring excellence and theological pre-eminence of the latter. Of these common elements, the 'deep' in Genesis 1:2 corresponds etymologically to Tiamat in the Babylonian myth; but whereas Tiamat was 'the goddess of the primeval World-Ocean, who had existed from time immemorial and was the mighty foe of the Creative God',[23] in Genesis God has no rival; the deep is merely part of the created world, waiting to receive whatever impress his will places upon it. Again, the ancient myths included stories of battles between the gods and great sea monsters such as the Dragon of the sea, Leviathan the Fleeing Serpent and Leviathan the Twisting Serpent.[24] In Genesis 1 the 'great sea monsters' are singled out deliberately (verse 21) and cut down to size as simply members of God's creation, the recipients of his blessing. Similarly, in a manner calculated (again probably deliberately) to deny any grounds for religious veneration[25] the sun, moon and stars are introduced as the Creator's handiwork (verses 14–18), the sun and moon being dignified not even by their common names, but only by their functions; and the sun, made on the fourth day, being pointedly distinguished from light, called forth on the first. These comments suggest very plausible reasons why the elements mentioned—the deep, the sea monsters and the heavenly bodies—are introduced into the narrative in the way they are. They betoken not a derivation of Israel's doctrine from the myths of surrounding nations,[26] but a pointed correcting of theologically-dangerous conceptions in the interest of a true understanding of God and his world. There is nothing here, in short, which precludes the idea that Genesis is God-given revelation, 'God preaching to men'.

We shall meet the subject of interpretation again, inevitably, in subsequent chapters, and also the question of myth; but a few further remarks of a general nature are appropriate before this chapter closes. Biblical scholars to-day often understand the first chapter of Genesis as 'priestly doctrine', gradually refined over the centuries and brought into its present form by an editor or redactor. The second and third chapters are more commonly regarded as myth, though serious and profound myth. These views, even if they were correct, would surely be profoundly inadequate. As I have argued earlier, the arrest, trial and crucifixion of Jesus Christ can be understood, in his own terms and to his complete satisfaction, by a purely secular historian; and his description of events and his explanation of causes may be proof against all arguments. But to faith such an understanding (taken alone) is hopelessly deficient. To faith (not in the sense of capacity to believe or propensity to do so, but in the sense of spiritual perception) this event is the turning point of universal history, the prodigious act in which God moved to reconcile a world-in-revolt to himself. Through the Bible the eye of faith has been opened to see it this way; it was God's act, not man's, however much man was implicated in it.[27] Now what I have maintained is that something similar is true of the Bible itself, and of the Genesis narratives in particular. They are God speaking, and it would be entirely inadequate to understand them *merely* as 'priestly doctrine'[28] and 'profound myth'. Critical scholarship has moved a long way in its appreciation of this wonderful literature since the Babylonian creation myths were first discovered (in 1853) and Charles Darwin published *The Origin of Species* (in 1859). It may still have much further to go; but the Christian reader will echo the affirmation of a great near-contemporary biblical scholar, Gerhard von Rad.[29] He was referring to the comment of a Jewish scholar, Franz Rosenzweig, that the sign 'R' (for the postulated 'redactor' of the Genesis documents) should be interpreted as Rabbenu ('our master'), a tribute to the greatness of his work. 'But for us,' says von Rad 'in respect to hermeneutics, even the redactor is not 'our master'. We receive the Old Testament from the hands of Jesus Christ, and therefore all exegesis of the Old Testament depends on whom one thinks Jesus Christ to be'. That is well said, and it is worth pondering.

In my work as a biologist researching into fundamental problems it fell to my lot to introduce radioactive tracers into plants; to kill and fix their delicate tissues in powerful chemical reagents; to embed them in tough epoxy resins; to cut them into incredibly thin slices (about a millionth of an inch thick); and finally to examine them in the electron microscope, at a magnification of anything up to 500,000. All this treatment the tissues, of almost infinite delicacy of structure, bore with great dignity. I often felt that my own dignity as a human investigator suffered much more, as very occasionally, with an exquisite pang, I realized that the scientific, analytical, manipulative approach had all but obliterated the marvel of the living thing. I sometimes wished, with scarcely bearable longing, to be able to recapture the wonder with which a child looks at a flower in its hand, or watches a tiny moving insect. Alas, the analytical approach, for all its necessity and worth, is potentially devastating. This has often been borne in to me as I sat in a gathering of professional biologists; rare indeed is it to find one who still regards his research material as 'fearfully and wonderfully made'.

Now what is true of the research biologist is, I am afraid, sometimes true also of the critical scholar. He dissects and analyses, compares and theorizes; and, before he knows where he is, the wonder of his material has gone. Of course he sees plenty to excite his interest and admiration (so does the probing biologist); but the glory has departed. 'Ichabod' has to be written over it. The Lord says, 'This is the man to whom I will look, he that is humble and contrite in spirit, and trembles at my word'.[30] There's the rub. Who retains the ability to tremble before something he has just analyzed into J, E, P and D?[31] Few indeed. And if the Bible is not that Word before which we must tremble, what is? Our society no longer stands in awe at the mystery of the foetus growing mysteriously in the womb. Haven't all its secrets have been probed and laid bare? So it is here. The human foetus and the divine Word: both have become cheap and commonplace in popular (and sometimes in learned) estimation. Without a tremor we abort the one and dismiss the other. No wonder our civilization reels.[32]

Important though the scholarly analytical approach may be, it is not all-important. What is all-important is the message God wishes the word of Scripture to bring to our generation. Critical scholarship has laboured to relate it to the age in which it first appeared, to the world-view then ascendant. What is needed is to relate it to the world-view now ascendant, that is, to the scientific world-view. This is our difficult but not impossible task. It must be done if our generation is to be persuaded to listen once more to the voice of 'God preaching' in these incomparable writings.[33]

CHAPTER FOUR

The Meaning of 'Creation' in the Bible

Creation (in the biblical sense) is not a theory alternative to, and in competition with, theories offered by science. It has nothing to do with materials, mechanism and process. It is an ultimate concept and thrusts itself upon us when we look, not backwards in time, but away from our space-time altogether. An analogy makes this plain, and also the relation between creation and providence.

God is entirely sovereign over his creation. Of his own will he brought it into being out of nothing, and by his energy he holds it in being. Thus he is both transcendent and immanent; another analogy illustrates this. Nature is therefore not divine; it is something over which man is to exercise responsible stewardship.

The word 'create' in our English Old Testaments is nearly always a translation of the Hebrew word *bārā'*. This latter is not, however, the only word used in connection with the subject we are discussing. There are, in fact, two other common ones: *'āśâ*, usually rendered 'make' or 'do' but with a wide range of other meanings;[1] and *yāṣar* usually translated 'form' or 'fashion'. But among these three words *bārā'* seems to be a term of special significance. It is used only of God (as is its Greek equivalent *ktizō* in the New Testament). Wherever it is used it seems strikingly to convey a sense of newness, sometimes indeed expressed quite explicitly.[2] It is probably true to say that *bārā'* (and *ktizō*) have a significance in the Bible very similar to that of 'create' in modern English, so it is worth looking into our own usage of this latter word a little more closely.

In current English we allow 'create' and its derivates to be used of human subjects as well as of God; we talk for instance of the 'creative arts'. But we still use the word in a special and rather exciting sense.[3] We remark that Beethoven 'created' the great *Missa Solemnis*; alternatively, of course, we may choose to say that he 'composed' or 'wrote' it. The two statements are not equivalent however; their impact is quite different. The one conjures up the picture of a man at a desk, pen in hand; the other, of a man with a rapt look in his eyes, deeply stirred by profound ideas. Again, it is quite appropriate to comment that another great musician, Haydn, composed a famous piece in the summerhouse at Eisenstadt while in the employ of the

34

Esterhazy family in the autumn of 1765; but it would sound a little odd if 'created' were substituted for 'composed' in this remark. 'Creation' is too powerful an idea to be tied down to a specific time, place and circumstance like this. In a sense Haydn's masterpiece was the work not of a few particular weeks but of a life; it came to birth not in a small summerhouse but in Haydn's world. It is this aspect of things that the word 'create' (as against 'compose') is used to express. 'Composition' is linked with time, place, circumstance and process; 'creation' soars above them all.

The biblical usage follows this pattern. Thus, when the Bible uses 'create' (or 'make' in a parallel sense) it is not concerned with mode and mechanism, or particularities of time and place.[4] The very furthest it goes in amplifying the mode of God's creative activity is to affirm, 'God said, "Let there be." ' Creation was by Divine *fiat*. 'By the word of the Lord the heavens were made', it says, 'their starry host by the breath of his mouth . . . For he spoke, and [the earth] came to be; he commanded, and it stood firm'.[5] The famous prologue to John's gospel[6] emphasizes the same with its striking metaphor of the 'Word' for Jesus Christ, the divine Agent of creation. Again, it is characteristic that when the biblical writers wish to say something[7] of the materials out of which, or the methods or processes by which things were given their present visible structure they avoid the word 'create'. They use instead the more general terms 'make', 'form' or others similar. Thus they tell us that man was 'formed' (not 'created') of dust from the ground, and woman was 'built up' (not 'created') out of part of Adam's side.[8] Again, Jeremiah was 'formed' in the womb; Job was 'made' and 'fashioned', and Jacob 'made' and 'formed' there.[9] The writer of Psalm 139 in an oft-quoted passage[10] uses four words in two verses to describe his development onwards from conception—but not one of these is 'created'. Accordingly it would seem fair to say that in the Bible the *creative* aspect of God's activity (and there are other aspects) is never linked to a particular time, place, process or material;[11] the act is seen rather as an unanalyzable movement out of the infinity of God's thoughts into the finiteness of time and space with all that fills them.[12] In this movement the only intermediary is God's Word. Creation is thus an ultimate concept; it indicates the furthest back it is possible to go in discussing the origin of things. It is in this sense that in John's vision of Jesus Christ our Lord speaks of himself as the 'beginning of the creation of God', the Word which gave effect to God the Father's creative thought and brought the world into being.[13] It is in this light that creation is seen, emphatically, not to be an alternative to scientific theories of origin. It is metaphysical, and moves (unlike the latter) in the realm of the invisible.[14]

However, it is important to recognize that in speaking of the notion of creation as the furthest back we can go we must not think exclusively, or even primarily, in terms of time (i.e. historical time). Why must we not do this? Because Scripture tells us that God's creative word is *still being spoken,* and will be to the end of the age.[15] It was being spoken, for example, in the great public drama of national regathering after the captivity of Babylon; and it was spoken in the private experiences of King David as he sorrowed for his sin.[16] Particularly where there is a notable new beginning, as in these cases, the Bible proclaims God as Creator. Even more significant for us are the references to recurring events in the living world. To each generation the call comes with immediate relevance, 'Remember also your Creator in the days of your youth'.[17] In every spring as the new leaves burst forth, or the new baby rabbits appear the Bible recognizes the creative activity of God. 'A new generation takes the place of the old. Creation continues, for God is perpetually sending forth His Spirit and renewing the face of the earth with fresh life.'[18]

Creation in the Bible is therefore an idea which comes into its own when we think not so much backward in time as upward (or outward) to God. In reflecting on the idea of ultimate origin the tendency to look backward in time has become natural and almost universal only because we and our world have so clearly the character of growing old. If we and it lacked this character (and we were still interested in the possibility of creation) we would be readier to grasp what has just been stated—that the essential direction to look is not backward in time but away to God. That our world *is* so clearly growing old acts powerfully to orientate our attention in a misleading direction. Nevertheless, this conclusion would seem to follow from the biblical handling of this topic: the idea of creation involves essentially a look in the direction of God, not of prehistory.

To clarify this further, let me use an analogy to illustrate the relationship of creation to another biblical concept, providence. Suppose we regard our world's on-going history as 'His Story', a story which the divine Author, unseen, is himself writing. His book is before us, incomplete; the ink is still wet where it lies open. Seeking to understand what is in process of being written we try reading what has gone before, turning back the pages one by one. The narrative grips us. Indeed we cannot pause until, alas! we find the writing becoming too faded, intermittent or archaic in style for us to decipher. At this point we stop; or rather, our interest, with less to hold it, becomes aroused in a new direction. 'Whose book is this?' we ask. 'Who wrote this story?' The question of authorship has met and challenged us as we travelled backwards to the beginning; but, in fact, this question could have confronted us at any point, even where the ink was still wet—only then we were too preoccupied with the tale itself. From any page we could have reorientated our thoughts, looked up, and exclaimed, 'Hey, who wrote this?' Better still, at the very start we could have called out, 'Is Anybody there, *writing* this?' For we notice, of course, that the story hasn't finished.

Though this analogy has imperfection (for example, we are not *readers* of God's story but characters in it) its interpretation should be plain. God, of course, is the Author of the tale, the Creator of the world it portrays and of the characters and encounters which enliven it. All is of his conceiving.[19] As we contemplate the story we note that fresh characters constantly appear, and new situations constantly develop. Where do they come from? At this point in our discussion the most obvious answer is, from the Author's fertile mind. This answer denotes creation; it explains, in fact, what is meant by that term. But the Author is not projecting into his created history a jumble of things and notions isolated, disparate, incommensurable, lacking relation and compatibility. He has thought (and written) into being a coherent story, in which everything is linked intelligibly and meaningfully with everything else. That fact introduces the possibility of a second answer not implied by mere logical necessity in the first: the characters and situations come from their antecedents *in the story,* and can be understood in terms of them. To the reader the *dramatis personae* and their circumstances do not appear suddenly from nowhere, as stones dropped into a pool might seem to do to a fish. They are engendered within the story itself in a way usually quite comprehensible to him.[20] Accordingly, as we trace with admiration the way the Author has skilfully (and lovingly) made one situation develop from another we find ourselves answering the question, 'Where did this come from?' in terms not of creation (as before) but in terms of what the religious believer, in real life, calls Providence, God's previous and concomitant activity in matters which influence the believer's present circumstances.

Seen in the light of this analogy the biblical concepts of creation and providence appear as two contrasting, but entirely consistent theological accounts of how a given situation arose. The first traces it from the conceiving mind of God *into* the arena of space-time; the second follows its 'history' *within* that arena. These accounts appear in relationship. They are not alternative to each other, as if we could opt for one or the other according to whim and fancy; they are complementary in the sense that we need both if we are not to miss important truths.[21] 'Remember now your Creator', and 'Honour your father and mother' is how the Bible reminds us of our present duty in both connections. One commandment recalls our *creation*; the other our procreation, i.e. the way in which God, in his story, has *providentially* woven our lives into the fabric of what has gone before.

The Bible's doctrine of Creation has been often summarized in the phrase *creatio ex nihilo,* creation out of nothing. The appositeness of this phrase should be apparent from the foregoing discussion, but some of its further implications need to be made explicit. Firstly, it means that nothing, visible or invisible, shares God's eternity. He alone 'has immortality'.[22] Without implying that eternity is simply time extended (in both directions) endlessly, this means that the Creator and his Creation stand on a fundamentally different footing. He is the Holy One[23] (as we saw in an earlier chapter), never to be confused with created things (even the most splendid, like the sun), or with the world process itself (which he controls).[24] The Bible never lets us lose sight of this distinction. The moment we do so, it says, we are in danger of falling under the power of a lie, great and damaging.[25]

Secondly, God is under no external constraints in his creative work. He hasn't to 'make do', like us, with whatever materials are available. He is absolutely free, bound only to be true to himself.[26] Of course, and most importantly, he didn't *have* to create the universe, or to create it just as it is. In other words, he designed it. This is a subject discussed more fully in Chapter 7.

Finally, as the Venite (Ps. 95) so magnificently describes, the world belongs to God, and is under his sovereign and omnipotent rule. Having created it, he is never, like the sorcerer's apprentice, faced with situations out of his control.[27] He commands, and creation obeys.[28] That man can (and does) resist God's will is, paradoxically, not a denial of his omnipotence but an outcome of it. It is he who has given this power to men. He holds them responsible for the use they make of it. That it is no denial of his omnipotence will be evident when God brings all eventually to judgement.[29]

We much now consider a little more closely a facet of the Bible's teaching hitherto noticed only in passing. The Deists[30] had made the mistake, biblically speaking, of regarding the world as a machine created to function on its own according to built-in laws. God was thought of as a sort of machine-minder who needed only rarely to 'intervene' to rescue the machine from a mis-function (a seized-up bearing shall we say), or to supply it with material. He exercised his control from outside. This, however, does scant justice to the richness of the Bible's teaching. This teaching is that God is not only 'over all' (transcendent) but 'in all' (immanent). He is not 'in' the material in the deep and full sense in which he is 'in' the spiritual, it is true; but he is in it in an important sense nevertheless. The relationship is perhaps something like that of a sponge in the sea. Each is 'in' the other (but not in an equivalent sense). Accordingly the Bible sometimes speaks of the creation being in God (the sponge in the sea); and sometimes of God being in the creation (the sea in the sponge).[31] There are two or three notable passages in the New Testament which stress this continuing dependence of the creation on God. Paul, speaking of our Lord as the divine Agent of creation, says, 'in him all things were created . . . and in him all things hold together', or cohere.[32] The great nineteenth-century scholar J. B. Lightfoot commenting on this says, 'He is the principle of cohesion in the Universe. He impresses upon creation that unity and solidarity which makes it a cosmos instead of a chaos. Thus (to take one instance) the action of gravitation . . . is an expression of His mind.'[33] No doubt Lightfoot would have agreed that Relativity and Quantum Physics are too. The writer to the Hebrews similarly makes of Christ the vast assertion, 'He reflects the glory of God . . . upholding the universe by his word of power'.[34] In a backward look at the course of world history the elders before the throne of God in the book of Revelation praise him because, 'thou didst create all things; by they will they were created and have their being.'[35] In the New Testament, whenever God's involvement with the dynamics of natural processes is mentioned (in general terms rather than with reference to particular historical instances), the verb is always in the present tense, which corresponds rather to our present continuous than to our present simple (e.g. 'I am doing' rather than 'I do'). Thus God *makes* his sun rise, *sends* rain, *feeds* the birds and *clothes* the lilies, our Lord tells us.[36] We miss his whole point if we regard all these statements merely as relaying beautiful thoughts. They are intended to disclose hard facts; there is little sense in making them if they don't. In the Old Testament it is the same. God's involvement with his creation is not something finished and done with when we reach Genesis 2. He *makes* springs gush forth; animal life of all kinds *looks* to him for food; stormy winds *fulfil* his command; he *gives* the fixed order of day and night, and so on.[37] He is the God in whose hand our breath *is*.[38] All these usages convey the same impression. God is very much involved in the continuing existence and functioning of his world; and it is again his *creative* word, Peter tells us, by which he gives effect to his will.[39]

The relationship we have been speaking about is, of course, what theologians refer to as God's immanence, his presence everywhere and in all things. This must be set alongside his transcendence, for the Bible, as we have seen, insists on both. This certainly poses a problem for thought; how are we to conceive of such a complex double relationship? It is natural in such circumstances to look for an analogy or model which embraces all its more important features. The Book and Author model we used before (and which we shall refer to again) served to illuminate the relationship of creation to time. What we want is one to illuminate particularly that of transcendence to immanence, and for this purpose I should like to adapt one suggested by Professor D. M. MacKay.[40] It runs somewhat as follows. Instead of thinking of creation as *manufacture* (the manipulation of matter into objects, like the Deist's watch) he suggests that we think of the process by which the busy two-dimensional world of a television screen, full of colour, life and interest, is held in being. The world on the screen has an existence of its own; yet it is clearly a dependent world. It is established by energy from outside; one throw of the switch and it dissolves into nothingness.[41] Moreover, its dependence is a moment-by-moment affair, and only the continuing goodwill of the broadcasting authorities keeps the show going. Yet this lively world need not be reflecting the goodwill of its sponsors. It may be filled with scenes of violence and hate totally abhorrent to them. Thus the model illustrates (however imperfectly) not only transcendence and immanence (as characteristics of the sponsoring authority, that is) but also that element of freedom to rebel which we know so well in our real world. However, as Professor MacKay himself warns, 'Every illustration brings with it a crop of possible misunderstandings which counterbalance its usefulness';[42] so we must not push it too far.

We may attempt to summarize the biblical teaching of God's immanence, so far as it is relevant to our subject, in this way. Nature is so constituted that it is legitimate to think of physical processes as taking place in accordance with fixed laws, laid down by the Creator and giving to Nature a certain in-built autonomy. Examples of this emphasis are Job 28:25,26 for atmospheric phenomena; Jer. 5:22 for terrestrial; Jer. 31:35,36 for celestial; Jer. 8:7 for zoological; and indeed the original ordinance of Gen. 1:11,12 which established for the biological world the fundamental principle that like reproduces like.[43] Yet we would be mistaken if we took this to imply that God's relation to Nature is like that of an aircraft pilot who has handed over his machine to automatic controls. This would be seriously to undervalue the Bible's teaching that it is God's ceaseless faithfulness and 'steadfast love' that alone maintain the law-abidingness of Nature.[44] Who would think it appropriate to speak of the faithfulness of a pilot whose plane flew automatically? The most he could mean would that the pilot was holding himself ready to assume control in an emergency—a quite inadequate representation of the situation of which the Bible speaks. Interestingly, the word *automatos*[45] (from which we get 'automatic') occurs twice in the New Testament (Mk.

4:28; Acts 12:10,11). On the first occasion it suggests the element of autonomy in Nature; on the second, it suggests the divine control over Nature. Evidently there is no conflict between the two ways of looking at things; but the whole tenor of Scripture tells us that the second is the more fundamental.[46] The ultimate truth is that whatever happens in Nature (whether through miraculous angelic mediation as in Acts 12:10,11 or in the ordinary course of things as in Mark 4:28), God ultimately, is the Giver of its actuality (if such an ugly phrase can be pardoned). When our Lord taught us that our heavenly Father 'feeds the birds and clothes the lilies', it is clearly unacceptable to interpret his meaning as if he had added, 'Circumstances permitting'. Jesus knew as well as we do that birds die in hard winters and flowers suffer malformations. His words are empty of power to reassure if they are qualified by any suggestion that in the end, God may be defeated by the vagaries of Nature. Nor can he possibly mean that, in such an eventuality, God like a good father will make it up to us in some other way. The only possible consistent understanding of his words is to take him as meaning, 'Don't worry; your Father controls *everything,* in wisdom, love and power.' And when he later added (in anything but the spirit of hyperbole), 'even the hairs of your head are all numbered', he indicated that this matter of the divine solicitude extends to the ultimately small. There would seem to be no excuse, on this testimony, for denying it even to the myriad electrons and photons that throng the universe of microphysics.[47]

For man, made in the image of God and charged with dominion under him, two serious corollaries follow from the Bible's teaching. The first is that what we have come to call Nature is not to be regarded as Divine, and worshipped. Men have often in the past worshipped the sun, moon and heavenly bodies, or certain animals and trees. Some still do. The Bible makes clear that this is a futile practice and a God-dishonouring error.[48] The truth is (the Bible implies), that as Nature was created out of nothing, so in the form we know it Nature will one day vanish into nothing, like a television picture when the power is cut off, or better, when the set is switched to a new channel.[49]

The second corollary is that man cannot escape his responsibility as the steward of the physical creation. It has become fashionable in some quarters to call the idea of man's lordship 'cosmic arrogance', as that able and influential writer Stephen J. Gould does.[50] This is foolish. The biblical teaching must be understood as a whole; and when it is, the denial of man's lordship over Nature becomes not cosmic humility (as Gould would imply) but cosmic escapism, dereliction of duty. For further, the Bible clearly teaches (what its critics overlook) that man's lordship is to be exercised on the pattern of his Maker's, for he is made in God's image;[51] and it leaves us in no doubt about this pattern. 'The Lord is good to all, and his compassion is over all that he has made'; 'Thou openest they hand, thou satisfiest the desire of every living thing'; 'He will gather the lambs in his arms . . . and gently lead those that are with young.'[52] Is then the charge of 'cosmic arrogance' against the Bible's doctrine of man's dominion over Nature justified? Certainly not. The lordship of man over the creation is to be expressed in compassionate stewardship, never in exploitation.[53] It is true that, like so many of Scripture's great principles, the mandate has been sadly misunderstood and misapplied by those to whom it was addressed; but this cannot excuse us from recalling and reemphasizing its real obligations.

The biblical doctrine of Creation is one of the distinctive glories of the Christian faith, and indeed of the faith of Israel from which it sprang. It is what the Scriptures (1 Timothy 4:6) call 'good doctrine', nourishing to the spirit and full of health and wholesomeness.[54] Its benefits are, in fact, unending. The man who receives it has at once an understanding of his humanity. He knows what he is, and where he is. And he knows why he is here. Gone are the spectres of meaninglessness, pointlessness[55] and absurdity. He hears the voice of conscience and duty. It may not be sweet and soothing, but it is no longer something trivial, arbitrary, irrelevant or unintelligible, a mere tormenting accident of existence. It sounds a challenge and points a direction. It gives a reason for behaviour and the promise of peace of heart. Or he feels the pull of romantic idealism, the unbearable longing for something 'beyond';[56] and it is no longer a mere will o' the wisp, a cruel deceit, an unaccountable self-deception of the human mind. If the Bible is true and God is Creator, however far off the object of desire is, it is at least *there,* as a Reality to be sought. Otherwise the future is undefined, undefinable, and arguably merely the stillness of universal death. Bertrand Russell, in his famous essay, *A Free Man's Worship,*[57] wrote that 'only on the firm foundation of unyielding despair can the soul's habitation henceforth be safely built.' That was in 1903; but this outlook, reinforced with the gloom of secular existentialism, still seems to hold in thrall many minds today. Against it the biblical doctrine of Creation stands as a sure defence. It is hardly too much to say that the choice lies between one and the other—biblical hope or secularist despair.

CHAPTER FIVE

Special Creation—and Chance

It cannot be maintained convincingly that the Bible teaches what is commonly understood by the term 'Special Creation'. It does, however, recognize the category of 'Chance' and itself uses it. But it does so in a way that in no sense limits its doctrine of universal divine providence. Even the throw of dice is divinely ordered.

This chapter will deal with two disparate topics, both, however, very relevant to the debate: Special Creation and Chance. We shall discuss them in order, and conclude with some general observations. First then, Special Creation: does the Bible teach Special Creation? The subject is an emotive one, for the words have been almost a battle cry for a certain school of conservative interpreters who believe—as I myself do—that all Scripture is 'God's word written' in the sense in which the Anglican Reformers used those words. One trouble is that the phrase 'Special Creation' is not always unambiguously defined. Because of this it can be a fruitless exercise debating whether or not it is a constituent part of the biblical revelation. Our first task therefore is to be clear as to its meaning.

Of course the Bible teaches that nothing is unloved, unpremeditated, without significance, merely accidental among God's creatures; in that sense all are Special Creations.[1] But that is not the meaning commonly intended. What is usually implied is that God *separately* and *individually* created many kinds of plants and animals on the appropriate 'day'. Once created, these bred essentially true (with very limited variability) in perpetuity. In specific terms therefore, oaks were created as oaks, horses as horses, and, of course, man as man, with no actual genetic connection between these different 'kinds'. To make matters clear, we may add that it is not implied that the 'kinds' do not share a common genetic code; that is not the type of genetic connection denied. It is what we mean by 'blood relationship' or common ancestry which is ruled out. However far we trace them back in time, it is proposed, we shall never find that the horse-line (for instance) joins on to the cat-line, or the human-line on to that of apes. Further, the lines will not change very much in themselves as we retrace them; they will retain their separate and

recognizable identity. This then defines what we shall mean by
'Special Creation'. As the multifarious sorts of animals and plants
were initially created, so essentially they remain, fixed.[2] Whether the
lines (when we have traced them back to their start) begin in adult
plants or animals, and in single pairs or in a larger group is not often
stated, except in the case of man. But it would seem necessary, in the
case at least of the higher animals, for them to have started-off as
adults.

We return to our question: does the Bible teach 'Special Creation'
in the sense we have given to that phrase? The positive evidence
adduced is twofold. Firstly, there are the general statements made
about the living world in Genesis 1:11,12,21,24,25. Secondly, there
are the particular descriptions given of the creation of man, in
Genesis 1:26,27 and Genesis 2:7. There is very little else in the Bible
which bears on the question. Occasionally Ecclesiastes 3·14 has been
cited in favour of the doctrine of the fixity of species, but most would
agree that this is as glaring a misapplication as is the citing of Psalm
93:1 in support of the idea that the sun moves round the earth. How
firm a biblical base have we then for the doctrine of 'Special
Creation'?

The short answer is: a far from firm one. In Genesis 1:11,12 the
earth is commanded to 'vegetate vegetation, herb yielding seed, fruit
tree making fruit after its kind'.[3] This literal translation brings out the
very general nature of the language, mechanistically considered. In
fact, it could hardly be more uncommitted. Besides the clear and
important implication that the earth and its fertility are not to be
deified but are subject to the creative command of God, the narrative
makes another very significant point. It was doubtless an early
observation that like reproduces like; that if one wants wheat, for
instance, one sows wheat seed, not seed of another sort. We take this
for granted, but, in fact, in the logical sense, the conclusion is not at
all a necessary one. Things could be imagined otherwise. That like
produces like is a *contingent* fact,[4] and contingent upon God's will,
the Bible implies. It is he who has decreed it should be so, and Paul
uses this truth to enforce a moral lesson in Galatians 6:7. It justifies
his point that it is *God* who is not mocked when men sow their wild
oats; men who do so are not just colliding with the laws of logic, but
with God's decree.

The creation of animals follows that of plants. It is in the same very
general terms, mechanistically speaking: 'Let the waters swarm with
swarms of living creatures', and, 'Let the earth bring forth living
creatures'.[5] In fact, it is difficult to see how the narrative could say
less than it does about what happened in terms of matter and process,
as the animal population established itself. Would it not be better to

rest satisfied with this reticence, and not to risk making it say what we feel it ought to say, rather that what it actually does?

Next we come to the phrase 'after its kind'.[6] It is in this crevice that the doctrine of 'Special Creation' is usually rooted. But how ample is the soil? How secure is the root hold? The phrase, as we shall see, occurs in a number of other contexts, and in them its meaning is usually clearer than it is here, where it can hardly be claimed to be self-evident. We must glance briefly first at its use in Genesis 1 and then at its uses elsewhere.

On its first occurrence in Genesis 1 (vv. 11,12) it could be taken as meaning that the 'seed' and the 'fruit' corresponded to the plants that bore them; this is the sense that has been remarked on above. This leaves the issue of 'Special Creation' quite untouched. All that it implies is that by divine decree, each plant reproduces its own kind, the sense being general and untechnical. However, the second group of occurrences (vv. 21,24,25) is a little different. No mention is made of reproductive mechanisms, so the above explanation may not apply—unless indeed we imagine the idea to be carried forward from v.11: 'God created every living creature . . . *to reproduce* after its kind'. A justification for this would be the subsequent command to be 'fruitful and multiply', but the suggestion may not commend itself to all, and it is safer to interpret the text without any reference to breeding. Interpretation therefore must involve paraphrasing the expression in a form conveying immediate sense, and this, in fact, is what the modern translations do in other places where it occurs. We pass on to these.

Consider, for instance, Genesis 6:19,20. The rsv[7] has 'Of the birds according to their kinds ['after their kind' rv] . . . two of every sort shall come in to you, to keep them alive.' The neb renders this 'two of every kind of bird . . .', and so does the niv. The jb is very close. Here the phrase has been taken to mean 'in all their varieties',[8] 'of every kind', with a stress on the number of different sorts. In Leviticus there is another set of occasions where the phrase is used similarly. Typical is Leviticus 11:13f. Of the birds that shall not be eaten the rsv has 'the eagle, the ossifrage, the osprey, the kite, the falcon according to its kind [after its kind rv], the raven according to its kind'. The neb has 'every kind of falcon, every kind of crow'; the niv 'any kind of black kite, any kind of raven'; the jb 'the several kinds of buzzard, all kinds of raven'. Here again the meaning has been construed as 'in all their varieties', 'all kinds of'. There seems no reason therefore why this meaning should not be read into the creation passages. If it were, the result would be immediately intelligible:

And God said, Let the earth bring forth living creatures of every kind, in all their varieties (Genesis 1:24).

One further use of the phrase occurs in Ezekiel 47:10. Here the prophet is speaking about the fish in the river which will one day issue from the Sanctuary and flow into the Dead Sea. The fish 'shall be after their kinds, as the fish of the great sea, exceeding many' (RV). The RSV translates this, 'of very many kinds, like the fish of the Great Sea'. The NEB has, 'every kind of fish shall be there in shoals, like the fish of the Great Sea'; the NIV is very close to the RSV; and the JB rather freer, 'the fish will be as varied and as plentiful as the fish of the Great Sea'. Clearly, the modern translators have taken the phrase to mean 'of every kind', with the 'every' clarified and amplified by the addition of 'as the fish of the Great Sea', or something similar.

If we accept that 'after their kinds' in the creation narrative can be fairly translated by 'of every kind' or 'in all their varieties' then we seem to be faced with the conclusion that we have a quite inadequate authority here for the doctrine of 'Special Creation'. For this doctrine to emerge clearly we would need to have the 'every' or 'all' particularized[9] by some such adjectives as 'subsequent' or 'continuing'. The passages would then read somewhat like this:

And God created . . . every living creature . . . of every *subsequent* kind, and every winged bird in *all their continuing varieties* (Genesis 1:21).

Were the record to run like this the doctrine of 'Special Creation' would be less in doubt; but this is just how it does not run. We may conclude therefore that the force of the phrase, 'after their kinds', is rather similar to that which is conveyed by the 'all . . . anything' in the related passage, John 1:3. It stresses the great variety, nothing less.

However, we still have the account of the creation of man. Is not he at least represented as originating in a separate and distinct act of creation, quite independently of the lower animals? Is not his 'coming into existence' described in terms which altogether deny any physical and biological continuity with them? I must confess that I quite fail to see that it is. Man was not, after all, allotted a special day of his own, but shared one with the land animals.[10] He was formed 'of dust from the ground'; but so were they.[11] He is certainly described as pre-eminent, as alone bearing the image of God. But this profound biblical affirmation cannot be pressed to mean a denial of physical and biological continuity. Man was surely no more unique among the commonalty of animals than was our Lord among the commonalty of men; and yet our Lord's uniqueness, so far from denying a physical and biological continuity with mankind, actually drew much of its significance from it.[12] He was 'born of woman', 'descended from David according to the flesh';[13] had it not been so he could not have been man's Redeemer.[14] In the light of the Bible's testimony that in

Jesus Christ 'the whole fulness of deity dwells bodily'[15] and yet that he was born of David's line, there seems little substance in the contention that man's being uniquely made in the image of God must imply discontinuity with other creatures, i.e. what we have agreed to call his 'Special Creation'. This doctrine of Special Creation therefore does not follow from the Bible's teaching about the nature of man, any more than it does from a consideration of the phrase 'after its kind'. The Bible leaves the matter quite open, as something of little concern to its main purpose.

We pass now from the discussion of Special Creation to that of Chance—or rather the biblical attitude to Chance. The importance of this may not be obvious at the moment, but it will become so later.

What is the biblical attitude to this tendentious notion? Well, first of all it must be recognized that it is a category which the Bible itself uses; it is regarded therefore as legitimate, at least in certain contexts. But it is important to see what these are. Ecclesiastes 9:11 is an illuminating verse in this connection. 'The race is not to the swift nor the battle to the strong'—necessarily or always that is—'but time and chance happen to them all.' Sometimes, in other words, the unexpected and unpredictable happens. Man never has total certainty in these things; an element of ignorance always remains. And it is this element of ignorance that is covered by the idea of chance. It should be noted that the writer is speaking of what can be 'observed', what is 'under the sun' (see NEB). In the invisible world of the spirit, therefore, 'time and chance' have no foothold; at least, the writer is not here giving them one. Incidentally, since 'observation' is the very basis of the scientific method, chance according to this verse, would be a perfectly valid concept for science. We know now, in fact, that it is an exceedingly valuable one.

Our Lord's use of the term in the Parable of the Good Samaritan (Luke 10:31) is similar. 'By chance', he says, 'a priest was going down that road.' The priest was taken by surprise; had he known the injured man lay there he would, of course, have altered his itinerary. Chance again is used to betoken ignorance; and that, in the last resort, is its only legitimate context.

Now man has had, ever since his primal act of disobedience, a sad but understandable reluctance to meet God.[16] However, he cannot help being a religious animal; so what does he do? He makes his own gods, of a sort which won't impose unacceptable demands on him and which he can manipulate. The mysterious and unknown, of course, must enter into their constitution, or they would hardly be gods. So he looks around for suitable material, and Chance suggests itself as an eligible candidate. It accordingly becomes deified, an active agency in its own right. Something of this sort is castigated by Isaiah as he thinks of the religious reverence given to Fortune and Destiny,[17] where Fortune is an idea not so very different from the one we are considering. But the thinking behind all this is futile self-deception, Isaiah declares,[18] and when God arises in judgement all such idols—Fortune, Destiny, Chance, Luck—will be swept away. They can't *do* anything,[19] and in the final reckoning they will be seen for what they are—non-entities.[20]

But the Bible goes further than this. While there are things *we* don't know, there is nothing God doesn't.[21] So there isn't really such a thing as chance for him. God is the master of all things, and he disposes even the throw of the dice.[22] For instance, when the people of Israel entered the Promised Land, it was divided to them by lot, but it was God who decided whose was what.[23] Thus Judah obtained Jerusalem, appropriately to the divine purpose that it should be the ruling tribe.[24]

Nowhere, perhaps, is the providential direction of chance events more powerfully asserted than in the story of the death of Ahab, an evil king of Israel. The story is told in 1 Kings 22. Ahab had seized the vineyard of Naboth after Jezebel his wife had procured the death of Naboth by stoning on a trumped-up charge. This brought to a head Ahab's long career of wickedness, and Elijah was sent to pronounce judgement against him (1 Kings 21:17f). For three years the sentence was delayed. Then Ahab joined his neighbour Jehoshaphat in an attempt to recapture Ramoth-gilead. He was an old hand at the art of war, and when the prophet Micaiah repeated the warning of Elijah, Ahab resolved to go into the battle disguised. The ruse was entirely successful; Jehoshaphat drew the enemy fire and the enemy turned his attention away from Ahab's sector. 'But' the biblical historian records, 'a certain man drew his bow at a venture ('at random', NEB) and struck the king of Israel between the scale armour and the breastplate'; and mortally wounded, Ahab withdrew from the field.[25] An obscure archer, a random arrow, a small area of vulnerability in the armour—what could more vividly convey the sense of the controlling providence of God? Ahab's final ignominious end was exactly as foretold by the prophet.[26]

Scripture is, in fact, full of cases where chance coincidences fulfilled God's purposes. Inasmuch as these were often foretold we cannot regard them as merely cases of opportunism on God's part; God's plan produced the occasion, not *vice-versa*. Thus Joseph was sold into Egypt;[27] Ruth became the progenitrix of Jesus;[28] and our Lord was born in Bethlehem.[29] A difference of a few days, or a few yards, in quite small events would have altered the course of history (often foretold history), very significantly.

To the Bible, therefore, the fact that an event can be spoken of legitimately in terms of chance, hap or randomness in no way removes it from the sphere of God's directing providence. 'The lot is cast into the lap but the decision is wholly from the Lord' (Proverbs 16:33). That is a truth the Bible would have us keep continually in mind.

Our conclusions therefore in this chapter can be easily expressed. 'Special Creation' is a doctrine that *cannot be established unequivocally* from the biblical data. That the Bible has been widely held to teach it is of no more significance than that Psalms 93:1 and 104:5 have been held to teach the motionlessness of the earth in space. A comparison with parallel passages where the Bible speaks of the righteous man or his foot[30] should have removed any grounds for dogmatism on this score! and the same is surely true over the question of 'Special Creation'. As a matter of fact, those who insist on the latter seriously weaken their case by distinguishing the created 'kinds' from present-day species. By doing so they at once open the door to genuine evolutionary speciation. Lion and tiger, two species: one 'kind'? Where does one draw the line? Is the whole cat family included in this 'kind'? What about the cheetah then?[31] and so on. The whole idea of the fixity of species—or kinds—seems to be due to an exegetical misunderstanding, comparable to that of the fixity of the earth.

Chance, on the other hand (the Bible allows), is a perfectly valid notion *within the context of human ignorance,* which is where, in fact, science finds it so valuable. Attempts to give it a wider validity than this and to imply that God has so constituted nature that even he does not know the outcome of ultimate events[32] (that is, that 'God plays dice'),[33] find no warrant in the Bible. It will be argued in a later chapter that the emphasis in Physics on chance and indeterminacy is not incompatible with the biblical view of all things under the hand of God. For the moment it needs only to be stressed that the 'chance' element in evolutionary theory presents no insuperable obstacle to biblical doctrine. This was realized by conservative theologians like B. B. Warfield and James Orr in the past.[34] There is no reason to modify their conclusion now.

CHAPTER SIX

The Primal Creation

The primal creation, though it was pronounced 'very good', was not an idyllic paradise. The great Adversary had access to it; it was perhaps designed to be the scene of his defeat. It was created as it is now, 'subject to futility', with elements of fear, predation, pain and death. Man was given the task of subduing it and bringing it to harmony. Through an act of disobedience at the beginning of human history he failed in his mission. Creation fell under the curse of disappointed hopes and broken relationships. It awaits man's final redemption and the uniting of all things in Jesus Christ.

It is common knowledge that the Bible teaches that after God's initial work of creation something tragic happened in the Garden of Eden: man disobeyed his Maker, and brought disaster on the race. The nature and extent of this disaster (according to the Bible's own testimony) we shall have to look into later; first we have a more immediate matter to consider—the character of the physical creation as it first left the hand of God. What was it like? Our conclusions about this primal creation (as we may call it) are going to be rather different from those which have commonly been held, so a brief statement may be helpful at the outset to set the matter in context.

Perhaps the commonest view of the primal creation is that expressed by Milton in *Paradise Lost*. Eden was, according to this, a Paradise, a place of ideal bliss. But even outside Eden (since the epithet 'very good' is applied to the whole creation in Genesis) there was nothing to 'hurt or destroy'.[1] Diet for both man and the animals was wholly vegetarian, and there was no pain, disease or fear among them. Perhaps there was even no death; certainly there wasn't for man himself. In the physical world eathquakes, droughts, volcanic eruptions, storms and such like were probably unknown. In fact, everything was idyllic—until man sinned. Then things changed dramatically for the worse. This is probably a fair statement of what has been held by many (and is still held by many) to be the Bible's teaching. Against this I shall argue that the primal creation was not idyllic; that some animals were predators, and that all were mortal; that even man himself was probably a flesh eater; and that storms and floods were as much a matter of course then as now. This to many will

50

be a fairly radical reinterpretation, and I shall have to set out carefully my reasons for adopting it. It would be disingenuous to maintain that it has no attractiveness as making a reconciliation with the scientific view easier. It obviously has. But it seems to me, nevertheless, to be the conclusion to which the Bible itself leads us. Let us examine the evidence.

The view that the primal creation was flawless is based principally on the repeated statements in Genesis 1 that the work of the successive days was 'good'.[2] We note that except for a minor variation on the first day, the formula which announces this is quite uniform and in the singular; 'it was good'. It seems reasonable to suppose, therefore, that it was the work done, the step taken, rather than the creatures made, to which the 'good' primarily applies. At the end, God saw all that he had done,[3] and 'it was very good'.[4] In the common view, this is effectively taken to mean that, judged from our present day standpoint, everything was there and then flawless; the whole contained no single source of disharmony. This conclusion is hardly beyond question, as we shall see.

A supporting argument is based on Genesis 1:29,30: 'Behold, I have given you every plant . . . for food.' On the face of it this appears to say that man and the animals at the beginning were wholly vegetarian. This conclusion is apparently reinforced by God's words to Noah when the earth was re-peopled after the flood: 'Every moving thing that lives shall be food for you; and as I gave you the green plants, I give you everything'.[5] This a new concession, it seems, made to man fallen from innocence. With it comes 'fear' and 'dread' of him into the animal world. Interpreted thus, this reference adds its support to the view under discussion. All-in-all therefore, it has been argued, the Bible gives us the picture of a primal creation where all was harmonious, and where the predatory habit, and fear and pain were unknown. The Messianic age it seems, will restore this, for then, 'they shall not hurt nor destroy in all my holy mountain'.[6]

Now the trouble with this view is that it jumps to conclusions far too quickly. It makes the assumption, for instance, that it knows exactly what the Bible means when it says that what God had made (or done) was 'good' or 'very good'. It tends to make these epithets absolute, instead of relative, and this is especially apt to mislead when the verb is translated *made*. But if they are taken as relative (as they surely must be, for God alone possesses absolute goodness)[7] a natural question is, relative to what? and an obvious answer is, relative to the purposes of the Creator. Before we consider the wider question this raises it is worth noting a minor parallel, also from the Pentateuch. The land of Canaan where Israel was to start its national life God pronounces 'a *good* land . . . flowing with milk and honey'.[8]

Yet in spite of this it was occupied by fierce aliens; it required both hard fighting to possess it and hard work to exploit it; and it remained surrounded by potential enemies. Its God-declared goodness did not reside in its splendid climate, superlative scenery, natural resources, freedom from threat and whatever else makes people happy; we could all probably think of places much better. It resided rather in its eminent suitability for God's purpose of blessing and training his chosen people.

What of the question then of the purposes of God in creating our universe? Here Milton is almost certainly on biblical ground in linking our destiny with other and non-material orders of creation. The Bible bears witness to the fact that our physical cosmos is not all that God has created. Our Lord often spoke of the angels, both good and bad;[9] Peter writes of 'angels, authorities and powers';[10] Paul of 'the principalities, the powers, the world rulers of this present darkness, the spiritual hosts of wickedness in the heavenly places'.[11] Of course, belief in such orders of created intelligences (particularly evil ones) is widely rejected today, especially in educated circles. Why? 'Theological and philosophical fashion' is the principal answer. There is no actual evidence against this belief; on the contrary there is a great deal of evidence for it.[12] The devil (to be particular) may be out of contemporary scholarly favour, but clearly he's not out of a job. Like all wreckers he prefers to work in secret, unrecognized and unsuspected, so the present climate of opinion is not surprising, and no doubt suits him well.[13]

Now the bearing of this on our subject should be obvious. The primal creation was one to which, in the wisdom of God, this great Adversary was permitted access.[14] Why, is not clearly revealed. What the Bible does tell us is that the work of Christ, conceived in heaven and wrought on earth, has reference to more than just the plight of humanity, however central to it that may be. 'He must reign till he has put all his enemies under his feet', Paul writes; and this includes Satan and the fallen angels.[15] In this work of overcoming angelic rebellion and handing over the kingdom to God the Father,[16] humanity has a central significance. It is 'through death' at the hands of men that our Lord 'destroys him that has the power of death, that is the devil';[17] and it is through his death, confessed by faith, that redeemed humanity itself overcomes the devil and participates in his defeat.[18] Thus, Paul says, it is 'through the church' (i.e. the body of believing human beings) that God the Creator makes known 'his manifold wisdom . . . to the principalities and powers in the heavenly places' as he fulfils his eternal purpose to unite all things in Christ Jesus our Lord.[19]

What all this suggests is that God had a purpose in the creation of our physical cosmos that reached beyond it, just as he had a purpose in calling Abraham that reached beyond his own race.[20] He already had a rebellion on his hands, and our world was to be the scene of an act (of supreme cost and self-giving) by which he would not only reconcile our world to himself (for that world would join the rebels) but also achieve the end of all rebellion and bring in everlasting righteousness.[21] If we accept this, it is bound to influence the view we are prepared to take of the primal creation. We shall hardly expect the latter to be a state of perfect bliss, an idyllic paradise. We shall rather be ready to understand the 'good' and 'very good' of Genesis 1 in terms of the stern (but loving) programme the Creator had in mind for his new creature, man. At this programme we must now look.

It is expressed in the mandate given to man in Genesis 1:28 which reads, 'Be fruitful and multiply, and fill the earth and subdue it; and have dominion . . . over every living thing . . .' This mandate thus charged man with 'subduing' the earth.[22] The Hebrew word for 'subdue' is *kābaš*, and in all its other occurrences in Scripture (about twelve in all) it is used as a term indicating strong action in the face of opposition, enmity or evil.[23] Thus, the land of Canaan was 'subdued' before Israel, though the Canaanites had chariots of iron;[24] weapons of war are 'subdued'; so are iniquities.[25] The word is never used in a mild sense. It indicates, I believe, that Adam was sent into a world where all was *not* sweetness and light, for in such a world what would there be to subdue? The animals, it suggests, included some that were wild and ferocious,[26] and Adam was charged to exercise a genuinely civilizing role and to promote harmony among them.[27] In fact, this function is set out very suggestively in Psalm 8, where man's Godlikeness, his strong delegated authority ('all things under his feet'), his encounter with opposition ('the enemy and the avenger') and the secret of success (the open celebration of God's glory, even by babes and infants) are the significant emphases. What man failed to do it fell to the lot of Jesus the Messiah to accomplish, and it is no surprise, therefore, to find this psalm referred to our Lord in the New Testament.[28] All this seems to justify us in believing that man's role was designed to be a Messianic one.[29]

We pass on to consider what is involved in the 'dominion' over the animal world with which man was charged. In common with 'subdue', the idea here is, as von Rad notes, surprisingly strong. It would seem indeed rather inappropriate if all man had to do was to exercise a gentle beneficence. What then did it involve? Permission to use animal flesh (as well as plants) for food? Calvin certainly had an open mind on this question.[30] I shall argue at once that it did, and then attempt finally to reconcile this viewpoint with the statements of Genesis 1:29 and 9:2,3.

Soon after the expulsion from Eden man was keeping sheep.[31] Indeed animals of a domesticable sort seem to have been explicitly included in the initial act of creation.[32] The occupations of Cain and Abel are introduced in a strictly parallel fashion; the presumption is therefore that their purpose was similar, in the main to provide food and clothing. Abel's sheep can hardly have been only for religious sacrifice, as has been urged; for Abel brought only the *firstlings* as an offering. Further, of those sacrificed only the 'fat portions' appear to have been burnt;[33] there is a strong presumption therefore that part of the sacrifice was eaten by the worshipper, (a practice to be regularized later in Israel's history). Again, sacrifice was not apparently a frequent event;[34] it can hardly therefore have been the main purpose of Abel's sheep-keeping. If this main purpose was to provide skins for clothing,[35] what happened to the carcases? They can hardly have been left to rot, for how then could they ever have come to be regarded as a thing worthy to be offered in sacrifice? The biblical record seems therefore to point to the conclusion that man was, at least as early as Abel, a regular eater of animal flesh, and not only so in connection with sacrifice. Can we go further back than that? Yes, possibly. We have the statement that immediately subsequent to the Fall the 'Lord God made for Adam and his wife garments of skins, and clothed them'. We are not to suppose that the Deity physically fabricated the garments; rather he gave the word of instruction as to what Adam and his wife were to do. But nothing is said about how the skins were to be acquired. Presumably this was by slaughter; but if this had been a radically new departure for man it is plausible to argue that this would have been made explicit. It is at least possible therefore that skins were ready to hand, having been used up to that point perhaps to construct shelters. This takes us back therefore to before the Fall, i.e. into the primal creation.

We turn to the New Testament evidence. There is an important reference to what was almost certainly the eating of flesh[36] in 1 Timothy 4:3,4. Paul is warning Timothy against those who 'forbid marriage and enjoin abstinence from foods which God created to be received with thanksgiving by those who believe and know the truth'. This is a strongly worded statement, with no obvious reference to the post-diluvial concession of Genesis 9:3. Rather, 'created'[37] takes us right back to Genesis 1, an impression reinforced by the next verse: 'for everything created by God is *good*', as that great chapter stresses. We may conclude therefore that Paul is referring to the primal order before man fell, and that man's 'dominion' then included the use of flesh for food.

A quite distinct argument concerns the practice of our Lord. *Even after his resurrection,* 'in the power of an indestructible life'[38] he himself partook of animal food and provided it for others.[39] Would this have been likely, it may reasonably be asked, if the eating of flesh had been a concession to man *as sinner*? Concerning marriage— coupled with foods by Paul in the passage we have just examined— our Lord did not hesitate to re-direct his disciples' obedience to the creation ordinance established 'in the time of man's innocency'.[40] If man was in the time of his innocency a pure vegetarian why did not our Lord re-direct his disciples to *this* ordinance too, instead of ignoring it? It would have been quite practicable (as experience both then and now confirms), and *ex hypothesi* probably beneficial. I believe the right answer is that pure vegetarianism was not a creation ordinance *in the sense in dispute*; that is, it did not represent the primal *status quo*.

There remain to be considered two outstanding passages in Paul's epistle to the Romans. The first is Romans 8:18–25, especially the statement in verse 20: 'the creation was subjected to futility,[41] not of its own will but by the will of him who subjected it in hope.' This statement is most often interpreted as referring to the curse of Genesis 3. Yet the whole passage has no noteworthy *verbal* affinity with the latter, whose memorable language Paul might so easily have taken up into his own rhetoric, in the manner he so often adopts.[42] It is true that there is an affinity of *ideas* between the two passages, Paul's picture of the 'whole creation groaning in travail' recalling the words of Eve of 'pain in child-bearing greatly multiplied', and death forming another link (if we interpret 'futility' as implying death). This we can agree. Yet I still feel a difficulty besides that of lack of obvious verbal affinity. To extract the whole range of animal sorrows— predation, savagery, fear and death with all else that could be read into 'futility'—from the simple terms of the curse (which mentions only 'the ground . . . thorns and thistles') is to go far beyond what the terms themselves suggest. It is surely sounder exegesis to limit the meaning of the curse as far as possible to what it actually seems to say, if this yields an adequate sense; and the sense that man's relationship with his natural environment was henceforth to be a blighted one is serious enough. We shall discuss it further below. What is being suggested therefore is that we should abandon the interpretation that equates the 'subjection to futility' simply with the Genesis curse, and undestand it instead as referring to the primal creation itself. Meanwhile, it is not necessary to dismiss as of no significance the presence of the two important ideas (*travail* and *death*) linking Paul's passage and the curse; they retain importance as part of the wider view.

There is, moreover, an objection of a different sort to the traditional interpretation of this great passage. To identify the 'subjection to futility' with the curse is to anchor it to a particular moment *within history;*[43] it is from this moment that Paul's thought is made to take off. But Paul does not seem to be in such a temporally-limited frame of mind in this chapter; witness verses 29 and 30, where the span of his thought is from foreknowledge (before history began) to glory (after it has ended). Is it not likely, we may ask, that the same is true of his thought in the passage we are discussing? If this is so, its span would then be from the very conceiving of the physical creation to its fulfilment in the 'liberty of the glory of the children of God'. On this understanding of Paul's words the 'subjection to futility' comes within the purview of Genesis 1 rather than of Genesis 3, which is what we have been maintaining.

The second passage is the moving peroration with which Paul closes the eleventh chapter of Romans:

> O the depth of the riches both of the wisdom and knowledge of God! how unsearchable are his judgements, and how inscrutable his ways! For who has known the mind of the Lord or who has been his counsellor? . . . For from him and through him and to him are all things. To him be glory for ever. Amen.

Of what is Paul thinking when he uses the words 'unsearchable' and 'inscrutable'? Principally, no doubt, of God's ways as Redeemer. But he can hardly be thinking exclusively of these. One of his principal lines is from Isaiah 40, a chapter rich in allusions to creation. Further, the 'from, through and to' of Romans 11:36 is too close to other similar Pauline passages[44] for it to be denied that God as Creator is also in view here. Now it is clearly inadmissable in the case of redemption to limit this great ascription of praise to a historical process which began only with Adam's sin (as if redemption was an *ad hoc* idea only then conceived). In terms of time it *must* span the whole of God's revealed activity (or more) from Genesis 1:1 onwards.[45] But if that is true for redemption it is *a priori* likely to be true also for creation. Again, Paul powerfully asserts that God's ways run utterly counter to what human wisdom would expect; 'God has consigned all men to disobedience that he may have mercy upon all.'[46] He is speaking here, of course, of God as Redeemer; but there is no reason to disbelieve that God's ways as Creator partake of this same character. They too are 'unsearchable' and 'inscrutable', and that from the very foundation of the world. They no more make sense to natural human wisdom than does God's plan of salvation. For this reason therefore I find no difficulty in believing that, in God's inscrutable wisdom, the animal world was *created* 'subject to futility'; that is, subject to the same imperfections as we see it to have now. Man failed in his mandate to lead it to liberty,[47] and now he is himself in thrall to futility and death and awaits his own final redemption. With his revealing in glory as God's son, Paul asserts, the animal creation too—and no doubt more beside—will attain its own glorious fulfilment, and its travail will be at an end. This is the cosmic hope to which it is looking forward.[48] If it be objected to this reading of the scriptural evidence that there is a grave moral difficulty in believing that God *created* animals subject to such evils as have been outlined, there is an immediate reply. Is it any less of a difficulty to believe that he would have subjected them afterwards to the same evils *through no fault of their own?*[49] Surely not; for the evils we are thinking of go far beyond what a mere solidarity between man and the animals would entail.[50] Thus the moral objection cancels itself out.

I would not wish to assert dogmatically that the view that has been advanced here *is* the Bible's teaching; but I believe that it is sound exegetically, and that it makes better sense than the usual view. Of course, as we noted earlier, it is easier than the latter to harmonize with the theory of organic evolution; but that should not be allowed to prejudice us either for it or against it. Nevertheless, *were the cases for both interpretations hermeneutically exactly equal,* evidence of an extra-biblical (i.e. scientific) nature should surely be allowed to influence the view we take. This principle has been universally accepted in connection with the interpretation of biblical passages which seem *prima facie* to imply the mechanical fixity of the earth (e.g. Psalms 93:1; 104:5), and the result has been a genuine deepening of our insight into Scripture. It is at least possible that the same may be true in connection with our understanding of the primal creation, and that here too Science has something to offer. All truth is God's truth; and as with meats, 'nothing is to be rejected if it is received with thanksgiving' (1 Timothy 4:4).

We may summarize the biblical picture of the primal earth as it was 'in the time of man's innocency' as follows:

1 It faced a brooding, antagonistic and personal 'Power of darkness', of whose origin the Bible tells us little.[51] It was destined to be the theatre of his overthrow, and this seems to be already implied in the very strong mandate given to man. The primal creation can hardly therefore be regarded as idyllic.

2 The Bible gives us no real reason to doubt that in its general physiognomy and its flora and fauna the primal earth was essentially the same as it is today; that is, that there were predators and herbivores, fruit trees and thistles, sunshine and storm, and much else that we now see.

3 The primal creation was nevertheless 'very good' in view of the purpose God had in prospect for it. Man was to play a key Messianic

role, and the outcome of his filial obedience (we may conjecture from reading the Bible more widely) would be a happy race of men and women filling the earth, living in complete harmony with one another and with their environment and revelling in the divine presence. The earth's physical turbulence would have been subdued for beneficial use; the animal creation civilized and brought into concord by the gentle elimination of discordant elements;[52] and the plant world encouraged into luxuriance and beauty. We are not to suppose that the Curse (which will be discussed later) altered the direction and thrust of man's endeavours to master the earth; it only soured and largely frustrated them. The mandate originally given to man was to do precisely what in his noblest moments he still seeks to do; eradicate everything hurtful, promote peace, organize plenty, and attain in himself a wondering comprehension of the world he lives in. The setting of the primal creation provided for this programme to go ahead with speed and success; and the far-reaching directive of Genesis 1:29,30 ('I have given you every plant yielding seed . . . and every tree with . . . fruit . . . and every green plant for food') is interpreted accordingly, and in harmony with other scriptures, not as what was there and then in force, but as a goal to be worked to. Had man lived in filial obedience all this delightful outcome would have been happily and no doubt speedily achieved.[53] That was why God saw 'all that he had done, and behold it was very good'. Alas—but that is another story.

The Genesis witness has far-reaching significance. Some great Eastern philosophies are world- and life-denying. Existence, they teach, is an evil thing. Something of their attitude spasmodically appears in Western society: 'Stop the world, I want to get off.' Against this the Bible is robustly world- and life-affirming. Creaturely existence is good, it insists. But that is not quite all. It is to get better! Hope, in the Bible, is one of the three outstanding forces in man's life. Significantly therefore, the present interpretation implies that it was in full exercise from the very first. Man was not brought on to the scene to enjoy a physical creation already perfect and needing no betterment. Rather, as the first creature qualified in full self-consciousness to be a fellow-worker with God, he was given the task of leading the animal world to a fuller liberty and more joyous life. This conclusion gains a great deal of its strength from understanding Paul's 'subjection to futility' not as a reference to the Curse, but rather to the work of the 'six days', Paul's 'eager expectation' (i.e. hope) thus assuming an earlier and more fundamental place in God's scheme of things.[54] That is surely as it should be, entirely appropriate for one of the three things which 'abide' (1 Corinthians 13:13).

CHAPTER SEVEN

The Garden of Eden

The story of Eden is to be understood historically. It is not 'myth' in the popular sense of that word. It is interpreted history, and takes us behind the scenes into the spiritual realm where human life finds its ultimate significance.

Man is a unity (though he has a dual nature); he is an 'inspirited body'. He began as an agriculturist. His naming of the animals anticipated an important activity of the modern scientist.

Man is the fully self-conscious animal. As such he was placed under the moral obligation of freely-chosen obedience to his Maker, a response which the lower animals were unqualified to render.

By himself man is incomplete; he needs a companion. So he was given woman to share his life in the most fundamental of human bonds. The meaning of this was revealed to him through the medium of a dream.

No incident in the Bible (except perhaps the story of Noah's Ark) excites ridicule, gentle or otherwise as the case may be, so readily as the account of Adam and Eve in the Garden. However, the laughter is loudest where the learning is least; to-day, the story of Eden is held in the greatest respect by many scholars of repute, even when they do not hold conservative evangelical views.[1] We shall devote two or three chapters to discussing it.

Interpretation

We have met this problem before, and we shall meet it again. How do we interpret the story? Do we regard it as myth, like the story of Orpheus and Eurydice? Do we read it as the story of Everyman in his moral experience? To do so, to be sure, would not be to take a contemptuously dismissive view of it, for while 'myth' (in popular parlance) can be a synonym for falsehood or sheer invention, it can also stand (in more scholarly circles) for the expression of profound truth, and as such is worthy of considerable respect. However, the fault with the view that the story of Eden is myth is twofold. First, myth and revelation (with which we are contrasting it) have entirely different functions. The function of revelation is to teach men to live in accordance with the will of God;[2] that of myth is to ensure that things continue in accordance with the will of man.[3] Second, while myth may powerfully express and illustrate a great truth it cannot *establish* it. It lacks authority, precisely that which is the foremost characteristic of revelation. On which side of this great divide the Bible stands is illustrated by a well-known incident. Our Lord was being questioned about divorce. He replied, basing his teaching on the story of the Garden of Eden: 'For this reason a man shall leave his father and mother and be joined to his wife . . . What therefore God has joined together let not man put asunder' (Mark 10:7,9). The accent on both function and authority is crystal-clear, pin-pointing the narrative (in our Lord's estimation) as divine revelation. Clearly, to him it was not myth, however elevated or profound. It was God expressing his will for man.

It is the same with Paul. He is grappling with the age-old problems of human sin, suffering and death; how could these terrible realities have achieved such a stranglehold over the lives of men? And his answer is firmly in terms of the disobedience of Adam in the garden. That was where man's slavery began, historically. Adam was no more a mythical figure to him than were Moses or Christ.[4] It has often been remarked that whatever may be said of other religions, biblical Christianity is rooted in real history, real geography, real events, real men and real women.[5] It is the most concrete and materialist of all faiths. That is why it makes the best sense, biblically, to read the story of Eden as real history. That is what I shall do here.

But granting that the Fall was historical, and a moment of fundamental significance in the development of mankind, it does not follow that the secular anthropologist working from his customary data and by his established methods, will find himself confronted at this point with a striking and inexplicable discontinuity. Scripture gives us no grounds for supposing such a thing. Everywhere in the Bible the great ethical and spiritual crises of humanity are such that their inner nature is hidden from secular eyes, and stands revealed only to faith.[6] Thus Abraham departed, like many another man, from the city of his birth; Moses tended ordinary sheep at the far side of the desert; our Lord was born in simple circumstances, lived as a carpenter, and was crucified as a common malefactor, almost unnoticed by the Graeco-Roman world.[7] These great events in mankind's redemption are recorded for us in Scripture as quite unspectacular happenings, the remoter ones leaving minimal evidence to serve as grist for the mill of historical enquiry. But minimal or not, Scripture nowhere implies or even allows that such evidence gives access to the inner meaning of the events recorded. That meaning is to be received on the authority of revelation alone.[8] There is no reason to believe that the Bible means us to regard the Fall as being any different in this respect from, say, the Birth of Bethlehem or the Resurrection at Jerusalem. The secular historian has no compelling reason for regarding either of these as falling outside the normal course of events; only faith knows otherwise. There is every reason, from a biblical point of view, to believe that the same is true of the Fall, a conclusion that must have an important effect on our understanding of the scriptural narrative.

Miracle

The biblical writers record many miraculous happenings, but they cannot be said to be avid for miracle. They don't see miracles lurking round every corner; they don't invoke them at every opportunity. Indeed, the careful reader must often get the impression that the opposite is the case.[9] The reason is a fundamental one. The teaching of the Bible is that God is as truly present in the ordinary as he is in the extraordinary; the difference is that he isn't so *obviously* or *significantly* present. Feeding the birds is as certainly a divine activity as feeding the five thousand.[10] Our Lord rebukes those who except they see signs and wonders will not believe;[11] and regards it as a sign of maturer faith to believe his plain word than to require the evidence of miracles.[12] The conclusion therefore to which the Bible is leading us is that we should recognize the hand of God in *everything,* and not just in the unusual or extraordinary.[13] Once we have grasped this great biblical principle and it has become part of our thinking we shall not be over-eager to attribute specifically miraculous status to things which the Bible tells us God did, or does, even when they are, as it were, one-off events.[14]

The creation narrative changes at Genesis 2:4 from what might be called the general to the particular. It was inevitable that it should do so in view of the purpose of Scripture, to direct conduct rather than to satisfy curiosity. This requires that man and his life should occupy the centre of the stage, and from now on they do. We prepare to enter the domain of human history. The change over is affected analogously to that in a well-presented natural history exhibit.[15] The background is a painted (i.e. symbolic) representation of sky, hills and trees; as we move forward this changes into real elements of plant, rock and soil surrounding the animal on display. The transition between the symbolic background (which gives a true impression, but isn't meant to be unimaginatively scrutinized) and the real foreground (which is to be understood literally, as it were) isn't by any means a sudden one. It is part of the skill of the arranger not indeed to hide it, but to make it unobtrusive. This helps the didactic purpose of the exhibit.

So the change is made from the symbolic extra-historical days of Genesis 1 to the common human days of chapters 4 onwards. A similar thing happens in the first chapter of John's Gospel, where we begin in eternity and end in common time, with something of both in between.

God formed man of the dust of the ground, a pre-scientific insight which we now know to be physically true, for man's body (here the subject) contains only the elements of inorganic nature. 'Forming' signifies a material process in space-time; it is significant that 'create' is not the word used.[16] To 'form' implies no particular mode of manipulation of the dust, for the word[17] is used of processes as diverse as growth in the womb and fabrication by woodworking and other methods.[18] Where the forming took place is not said, but the natural sense of the passage would suggest that it was not in the Garden. Man was placed there, after he was formed.[19]

Eden is evidently to be understood, from its association with well-known topographical features, to be a real geographical region, though the data given do not make possible an exact location. Within this region 'the Lord God planted a garden', an area designed to be both beautiful and productive. Here grew trees for man's higher satisfaction, his nourishment, his health (see the reference to the tree of life in Revelation 22:2), and, significantly, his moral and spiritual discipline (the forbidden tree).[20] Nothing here is scientifically absurd, philosophically inept or humanistically trivial; rather, scholarly commentators agree that everything is profound, theologically and psychologically. With the rest of the description of man's life in the Garden it constitutes a commentary on that universally-respected verdict of Deuteronomy 8:3 and Matthew 4:4, that 'Man shall not live by bread alone, but by every word that proceeds from the mouth of God.' Of course, the biblical writer does not expect us to understand the statement that 'the Lord God planted a garden' as implying a physical activity of the Deity observable as such,[21] nor is it likely that he expected us to interpret the language about the cherubim and the flaming sword as plain statement of physically observable fact. The 'forbidden tree' may well be symbolic of a prohibition about which for reasons of brevity and force the writer chooses not to be more particular.[22] These are not difficult points to appreciate; they employ literary devices common and indeed inevitable in all great literature dealing with profound matters. Of course, the anthropologist, the student of civilization and the secular historian all have their own, quite other, ways of speaking of these things, and we have no right to quarrel with them (nor they with each other, for that matter), so long as none claims exclusive validity for what he says. The Bible's specific function is different from all others; it is to take us behind the scenes into the realm of the unseen.[23] It discloses that however tightly observable events link up with one another in an apparently unbroken nexus of cause and effect, the whole drama of existence has an Author who is writing it even as it unfolds. As our Lord taught us, he daily 'makes his sun rise', and 'sends his rain'.[24] Nothing is more destructive of the whole purpose of Scripture than to label such language as this (common both to Genesis and to our Lord) as simply 'the Hebrew way of looking at things', and to dismiss it accordingly from serious attention. As we noted earlier, the Bible teaches that natural events are as much God's doing as those we choose to call supernatural. Correspondingly, the narrative we are considering does not imply that the events it describes were such that a hypothetical onlooker would have had to exclaim, 'A miracle!'. To insist that it does is to deny rather than to emphasize the majestic sweep of the biblical teaching on God's all-embracing providence. For he holds the 'whole wide world in his hands'; the roll of the dice as well as the fall of the sparrow is subject to his will.[25] 'Even the hairs of your head are all numbered' our Lord bid us remember;[26] the most trivial circumstances of our lives are watched-over and ordered by God's all-embracing providence. This may well be a biblical doctrine the reader feels he cannot accept. No matter; its importance at the moment lies in the light it sheds on the assertion that was made earlier: that the Bible teaches that common events are just as much God's doing as those we call miraculous, (the only difference being that they are less obviously and significantly so). Because of this a specific biblical statement attributing a happening to God cannot, *ipso facto*, be taken as asserting its miraculous nature. Unless the Bible is more explicit, the happening may well fit into the category the Victorian scientists called 'uniformitarian'.[27] Recognition of this is highly important if we are to avoid needless difficulties.

The account we are given of man's origin in the second chapter of Genesis indicates that human life of the kind we know began in connection with agriculture; he was placed in a garden to 'till it and keep it'.[28] This is an emphasis with a distinctly modern ring. However, in fulfilment of his mandate to 'replenish the earth and subdue it' his activity was not, it would appear, confined to the Garden, for the riches of the surrounding region are noted pointedly in Genesis 2:11,12. This suggests a concern with technology and art (furnished from outside) as well. The bringing of the animals to Adam for naming was also highly significant. It was his initiation into dominion over them by way of knowledge and reflection. For thought, in the sense of understanding, is helpless without words; it can't get moving. It has no tools with which to set about its task. It has first to develop a language in which to articulate itself and to communicate itself. And the first component of language is the noun or name. It is a very superficial and mistaken view to think that this noble and dignified chapter is merely purveying a pretty story. The opening words of John's Gospel ('In the beginning was the Word') are a pregnant statement about Jesus Christ as God and Creator. The metaphor they employ (*logos,* the Word) derives its value from something near to our familiar experience: that words constitute the very basis of human knowledge. Adam was no more exercising himself in pleasantries when he named the animals than is today's electron microscopist when he names the organelles he finds in his micrographs, or today's nuclear physicist the elementary particles in his bubble chamber. To those men of science who know the Bible it is the same God who brought the animals to Adam who still, in a manner fundamentally no different, brings these less commonplace, more esoteric objects to the attention of themselves and their scientific colleagues. This is one implication of what the Scriptures are saying at this point. God is the Teacher who instructs man about matters on the scientific level as well as about things on the moral and spiritual.[29]

It is in the Garden of Eden that Adam appears as the fully self-conscious animal. Whether or not other animals have rudimentary self-consciousness is a moot point,[30] but self-consciousness is of the very essence of human nature. It is noteworthy that in Genesis 1 when God had created the first animals he 'blessed them saying, Be fruitful . . .'; but when he had created man he 'blessed them and God said *to them,* "be fruitful . . ." '.[31] The personal address, implying the ability to comprehend, is surely significant. But the matter goes much further in Genesis 2, where God lays a moral obligation on man of which the animals know nothing 'You may freely eat of every tree . . . but of the tree of the knowledge of good and evil you shall not eat . . . '.[32] The idea of moral obligation, of duty, presupposes that man knows himself as himself, which the lower animals do not. In no other sphere, surely, does self-consciousness play such an important part as in this.

Before we pass on to consider the creation of woman there is another point of biblical teaching to note. Nowhere else in the Bible is there such a concise statement of the constitution of human nature as in Genesis 2:7. The description of man's creation is continuous, with two movements: the 'Lord God formed man of dust from the ground', and 'breathed into his nostrils the breath (*nešāmâ*) of life.' In this way man 'became a living being' (*nepeš*; soul, RV). Man, in other words, is not a soul imprisoned in a body, but a complex unity. He is an 'animated' or 'inspirited' body.[33] This view is remarkable by comparison with other ancient teaching which denigrated the body. Again, it has a modern ring.[34]

In the Garden man also became conscious of an incompleteness in himself. As he got to know the animals it became apparent that there was no 'partner' (NEB), no helper 'opposite to' him among them, no colleague as we might say today. In these terms the Bible emphasizes first, that man has a demanding work to do ('helper' highlights this); and second, that woman has a complementary part to play in it. Replenishing the earth, subduing it, and exercising a dominion under God over the animal creation are no doubt all part of this task, which is given as a joint enterprise (Genesis 1:28). It should hardly be necessary to add that with attention in this passage focused on the task the designation 'helper' for woman no more implies inferiority than it does divinity (compare e.g. Psalm 146:5, where, in a common usage, the word is the same). Delivered from the curse, Scripture sees husband and wife as co-workers (*sunergoi,* Romans 16:3) and co-heirs (*sunklēronomoi,* 1 Peter 3:7).

The description of the creation of woman has usually been taken literally. To those who regard the whole story as a myth it presents no difficulty. Those who regard it as history, as I do, face a problem. They usually interpret it as a miracle (not of creation *ex nihilo* but of 'building into'[35] from a rib or side of Adam), the deep sleep being sent to Adam so that he should feel no pain. This apparently was Calvin's view. It leaves one with the difficulty that one miracle suggests another: why couldn't the operation have been painless *without* the deep sleep? Wouldn't Adam have learned more, and been more deeply impressed, if he had observed it all, as some commentators (e.g. Cassuto) suggest he may have done?[36]

I believe there is a better understanding of the narrative. From what was said earlier about miracle it would appear that it is sounder exegesis not to invoke the category unless the biblical data compel us to. Flying to it too readily and often undermines the significance of miracle. It cheapens it. What alternative understanding is there then, consistent with loyalty to the Bible?

A legitimate one, I believe, is that God used the medium of a dream to convey to Adam the true status of the woman he was to be given as wife. This, not anaesthesia, is the import of the statement about the deep God-sent sleep. Dreams can be very vivid and particular, and their memory very persistent and powerful. They can, in fact, have a life-long influence, as many people have testified. Dreams are often associated with deeply-felt longings and take the form of idealized fulfilments of them.[37] Things can happen in dreams which cannot be expressed satisfactorily in words. The dreamer awakes with his mind very definitely orientated and sensitized. All of these points make sense of the idea that God used a dream both to instruct and to prepare Adam psychologically for the gift he was about to make.

There is biblical support for this view. Three examples must suffice. Jacob had a revelation of very considerable significance in a dream.[38] So did Nebuchadnezzar, who learnt in a dream the real nature and future history of four great world empires of which his was one.[39] Even more instructive is the case of Solomon. In 2 Chronicles 1 we are given an account of how God appeared to him and invited him to make a request. Solomon asked for wisdom to fulfil the duties of kingship well, and the prayer was answered. There is no mention of a dream; the only possible hint is that it was 'by night'. That the transaction was real the Bible clearly intends to convey, for it records that the request was fulfilled in Solomon's subsequent administration. There is a parallel version of the story in 1 Kings 3 from which we learn in addition that the incident *occurred in a dream*. The Chronicles account can be aligned with the passage in Genesis, with its 'by night' fulfilling the same semantic function as the latter's 'deep sleep'. The making of the woman in Genesis is paralleled by the divine visitation in Chronicles. So far so good. But the twin passage in 1 Kings 3 enables us to interpret the 'by night' of Chronicles as 'in a dream by night'. I believe this goes a considerable way towards justifying the interpretation of the 'deep sleep' of Genesis 2 as 'in a dream during deep sleep'. It would, of course, mean that the event was not miraculous in the *popular* sense. But to Adam it would have been a word divinely spoken, by which he and his wife were to live.[40]

This interpretation seems to be true to the data and to the genius of Scripture, and certainly robs the account of none of its power or impact. Adam woke from his sleep to find the woman beside him, and with a God-given understanding of who and what she was, and from whom she had come. Psychologically and emotionally—such is the momentum of a vivid dream—he would have been prepared for her. Few methods of instruction could have stamped things so indelibly on Adam's consciousness, or linked two lives so indissolubly as a result.

No finer exposition of the meaning of all this—none truer to the divine intention—has ever been given than that by the old puritan Matthew Henry. 'Woman was made of a rib out of the side of Adam; not made out of his head to top him, not out of his feet to be trampled upon by him, but out of his side to be equal with him, under his arm to be protected, and near his heart to be beloved.' Further, he says, there was clearly a progression in creation. 'If man is the head, she is the crown. . . . The man was dust refined, but the woman was dust double-refined, one remove further from the earth.'[41] This chapter reveals the divinely-ordained pattern for the sexes, one which I believe the great majority of thinking men and women will agree could not be improved on. Adam is established in the headship, but with such balance and delicacy that we have here neither grounds for feminism, nor excuse for male chauvinism. I venture to think that this biblical teaching is one to which every true man and woman, unsubjugated by whatever transient spirit of the age holds sway, will spontaneously subscribe.

CHAPTER EIGHT

The Temptation, the Fall and the Curse

Man as a creature was necessarily placed under the obligation of obedience to his Maker. His temptation and fall are to be regarded as historical events; a merely psychological interpretation is inadequate. The New Testament unmasks the serpent as the great Adversary, Satan. The punishment for disobedience was death, the ruination of relationships—man with God, man with his fellows, and man with nature.

Nowhere does the Bible's account of man show its total superiority over that of secular evolutionism so strikingly as in the matter of moral experience. What is the nature of moral obligation? Why are its demands so clamant and inescapable? And why so comfortless and tormenting? Why does it turn us all into hypocrites, hard on others and soft on ourselves? Secular theory has no answer; the Bible's answer is both rational and satisfying, humbling and hope-bringing. Man's life is too often 'solitary, poor, nasty, brutish and short' (as Hobbes put it) for a very understandable reason. He is a rebel, alienated from God. Cut off from his true life, he is out of harmony with himself and with his fellows. That is the Bible's answer.

Let me illustrate. A beneficent ruler, wishing to make his people happy, constructs for them a system of roads, and presents them each with a splendid automobile. 'Cheers!' they cry as they take delivery. But cheers will soon change to tears unless, at the same time, their benefactor imposes on them a code of conduct, a highway code. Without this his beneficence will be worse than self-defeating, clearly a recipe for disaster. Even if his subjects share his altruism this will be so; and if by some chance any of them develops traits of self-centredness it will be more than ever necessary that his authority shall be there, publicly recognized in advance, to settle disputes early before they get out-of-hand.

This illustrates the rationale (partly, at any rate) behind the biblical statement that the Creator placed man under a moral restraint. The restraint was very limited compared with the associated liberty; but it was a restraint nevertheless. 'You may freely eat of every tree of the garden; but of the tree of knowledge of good and evil you shall not eat.'[1] It is customary in many quarters to scoff at the biblical narrative

68

of the Fall, but this is foolish. The profound theological and psychological insights of the story have often been recognized.[2] How subtle is the approach of the Tempter; how artless and uncontrived Eve's declension into self-pleasing; how fundamental and far reaching the results for the race! Loss of inner harmony is immediate. Then come fear, alienation from their Benefactor, disloyalty to a bosom companion, and finally murderous hate towards a brother.[3] All this is so perennially up-to-date and true to experience that the story of the Fall is with justice regarded as the story of Everyman. But this by no means exhausts its trueness. As the Cross was a unique historical event, and yet is the story of Every-disciple[4]; so (it may be convincingly argued) was the Fall a unique historical event, and yet is the story of Everyman. The New Testament, in fact, makes much of the parallel between the two, as well it might; for the correspondence between them (sometimes of similarity and sometimes of contrast) is too far-reaching to be accidental.[5]

This view (that the Fall was an historical event) raises some difficult questions of its own, however. Was the serpent a 'real' snake? Did the tempting constitute a visible scenario? What was the nature of the sinful act? What are we to understand by the 'cherubim and the flaming sword'? and so on. In a sense these are not very difficult questions to answer. Once we have grasped the biblical teaching that there are orders of reality beyond the physical; that our material world is, as it were, embedded in a world of higher dimensionality whose occupants can and do on occasion break audibly and visibly into ours—these matters cease to be a real problem for thought. They take their place easily alongside much else of which the Bible tells us—God's personal manifestation in theophany; the angel with the drawn sword who appeared to Balaam's ass; Elisha's chariots of fire; the angel of the Annunciation; the Holy Spirit descending in bodily form as a dove or as tongues of fire; the revelation on the Damascus road; and so on.[6] The situation is seen to have a self-consistent logic of its own to which science can say neither 'Yes' or 'No'. This needs to be clearly understood. These 'intrusions' from a higher world we have no reason to expect to be subject to our beck and call; and it is precisely this circumstance that places them firmly and in principle beyond the purview of science.[7] But that they are real is borne out not only by the Bible but by the abundant testimony of all ages, not excluding the present. It is idle to deny them, at least as life-changing experiences.[8]

Was there then a visible scenario? Many commentators hold the view that everything took place in Eve's mind; that she was confronted by the craftiness lurking in her own heart.[9] What we hear is a soliloquy, two sides of Eve's nature engaged in dialogue, the

serpent merely objectifying the dark one. Similar views have been
put forward about our Lord's temptation in the wilderness. Everything
is reduced to psychology. Now the most one can say for such ideas is
that they may be true descriptively of temptation as an experience;
the sinner is rarely (if ever) conscious of an objective Tempter. But
that is not to say that he is not (on occasion)[10] there. The saint is
rarely conscious of the Holy Spirit as an objective Helper, either. The
logic which would deny the Tempter therefore will find it hard to
affirm the Helper. The whole argument turns, in fact, in one way or
another on what is the main presupposition of the Bible: there is a
reality beyond this material world. There are things that are seen,
and there are things that are unseen, and the latter are the more
fundamental. The 'psychological' view owes far too much, in fact, to
the persuasion that science is the measure of all things. Granted an
unseen world, it is not difficult to believe that Satan entered into a
literal serpent as he entered into Judas (and even Peter), or the
demons into the Gadarene swine.[11] The sinful act itself (i.e. the
taking and eating) may be understood literally or as metaphor;[12] the
'cherubim and flaming sword' in the light of Elisha's horses and
chariots of fire (2 Kings 6:17) or the protecting angelic hosts (Psalm
34:7). Further than this our present apologetics need hardly go.

We approach now the question of the Curse. That man's life is
lived under a dark shadow hardly needs arguing. At the end death
takes all; but even before that there is pain, disappointment, lack of
fulfilment, hunger, conflict and misery. So much is this so for the
greater part of humanity that it constitutes a never-exhausted
argument against a God of love. It is an argument not easy to
counter, for there are times when even the most fortunate feel the
force of Omar Khayyam's sentiments:

> Ah Love! could Thou and I with Fate conspire
> To grasp this sorry scheme of things entire,
> Would we not shatter it to bits—and then
> Remould it nearer to the heart's desire?

Why should human existence be like this? Secular humanism has
not even an apology for an answer. How could it, when natural
selection preserves only what has survival value? Even if it could
think up some recondite non-Darwinian mechanism it is hardly likely
that it would measure up to that which it was invented to explain: a
feature so universal, so massive, so tragic and so essentially *ethical*. It
is here that the Bible comes into its own. It is the inevitable result, it
says, of man's alienation from God, and of the curse that fell upon
Adam. What was this curse? In one word, it was death.

In order to understand what the Bible means by death certain things must be made clear. In the first place, the Bible makes no suggestion whatever that when man was first created his earthly existence was to be endless. With a finite earth and the command, 'Be fruitful and multiply', it is almost self-evident that this could not be so. How his earthly life would have terminated we do not know; the nearest the Bible comes to telling us is the comment it makes about Enoch, that 'he was taken up so that he should not see death . . . God took him.'[13] Further, there is reason to believe that the function of the Tree of Life 'in the midst of the garden' was to ward off disease and old age,[14] and this again speaks against an inherent physical immortality. In agreement with this is the great messianic prophecy of Isaiah 65. When God creates his 'new heavens and new earth' (verse 17) in which 'nothing hurts or destroys' (verse 25), human life will still have its allotted span (verse 20) and then presumably be taken to a higher sphere.

In the second place we must suppose (following the argument of chapter 6) that Adam and Eve would have observed death in the animal world. The sight of a dead animal would, in fact, have exemplified the fate in store for them if they disobeyed. How exactly would the spectacle have struck them? Our scientific culture has conditioned us to think of death as a physical change of the thing-in-itself. Accordingly we try to establish death by instrumental tests.[15] But this misses the point. The biblical understanding of life connects it with knowing—existential knowing.[16] It thus implies entering into relationship—with God, with other persons and, to a lesser extent, with things. That is why the Bible links life with *light* (which facilitates relationship) and with *love* (which develops it).[17] To the Bible therefore life is not a property of the thing-in-isolation. It consists in cognitive and responsive relationships with things and especially with persons; and death is the ruination of those relationships. The most striking and poignant thing about a dead animal or person is not that it has changed-in-itself, but that it has changed-in-its-relationship to us and to the world. It no longer recognizes us, or makes any response to us. The sun comes out, and it shews no pleasure at the warmth of its rays, and so on. So far as others are concerned, and the great world outside, it knows nothing and answers nothing. That is the fate with which Adam and Eve are threatened, and this is probably how they would have regarded it. Relationship is the key category, both here and in the whole of Genesis 3 (and beyond). This can hardly be over-emphasized.[18]

'In the day you eat of it you shall die.' The opening phrase here is not necessarily a precise time-pointer; in the original Hebrew it is exactly the same as the corresponding phrase in Genesis 2:4, where it is indefinite in meaning. However, it is clearly unacceptable (if only because we have moved now into the realm of human history) to demand that it span the 900 or so remaining years of Adam's life (Genesis 5:5), as we must do if we insist that *physical* death is the primary reference. There is no doubt that the punishment for sin included physical death (*as we now know it*); for the event we now know as death has a sting in it, due to sin.[19] But (it is worth reiterating) we are certainly not free to conclude that before his disobedience man's earthly life was of an immortal quality and would have been endless. Death as penalty highlights the significance of relationship; and the most important of all relationships, the one from which all others are nourished, that with the 'faithful Creator', was shattered the very moment the forbidden fruit touched man's lips. Spiritual necrosis sets in instantly. He is 'cast forth as a branch and withers' (John 15:6); and ever since, he has existed (except where God reaches down in grace and rescues him) in the darkness of spiritual death. The New Testament especially speaks of this with great force and plainness of speech.[20] Other consequences followed inevitably and in their own time. Driven out of the favoured habitat of Eden, barred from access to the Tree of Life[21] whose healing virtues had warded off accident, illness or senility, man's pilgrimage became one 'long march into the night'.[22] Outwardly, human society became corrupt and estranged from nature, which from now on was something to be plundered and exploited; inwardly man's *psyche* became an arena where impulses of lust, greed and cruelty fight incessantly (and often successfully) with nobler instincts for the mastery. Finally, with no victory in sight, his 'years come to an end like a sigh'.[23]

There is a sense of realism about all this. It rings true; human life is like this. God is to most people the Great Uncertainty, more often than not regarded with very mixed feelings. People neither know him, nor (if they are honest) are they sure that they want to. 'Alienated from the life of God', 'estranged and hostile in mind' is how the New Testament describes man's present condition in relation to God,[24] a condition for which Christian theology has coined the phrase 'original sin'. Doesn't universal human experience corroborate all this? Surprise reminders of God do not commonly fill men and women with delight, gratitude and a happy sense of security. Alas, it is rather the reverse, Man is in the far country.[25]

It is in terms of a spiritually-changed relationship too that man and woman henceforth had to face life together. For Eve, to whom the experience of childbearing was to become much more painful, the clue is to be found not in deleterious anatomical or physiological changes but in the entrance of anxiety and fear.[26] Who does not know how easily these can change what might otherwise be an exhilarating challenge into a terrifying ordeal? Eveyday life is full of examples, from public speaking to driving tests. Then there entered into human consciousness a strange and discordant new emotion—shame. Man and woman became 'ashamed of their own naturalness', as Carl von Weizsäcker the distinguished astrophysicist, once put it.[27] Even now this remains a burden. Any attempt to deny it means brazening things out in defiance of an inborn instinct, with loss of poise and dignity. Sexual relationships became soured after the Fall. Tyranny now adulterates tenderness, sensuality sensitivity; love-hate is too often a frequent alliance in the encounter of man and woman. The Bible poignantly illustrates this in the moving story of the passion of Amnon for his beautiful and virtuous half-sister Tamar:[28]

> Then Amnon hated her with very great hatred. And Amnon said to her, 'Arise, be gone!'

'Yet your desire shall be for your husband, and he shall rule over you.'[29] Feminism may be a protest at this, but it tends to be a hard, loveless and self-defeating one. The 'gay' life may seem to some to offer a way round it, but it brings its own bondage and disillusionment.[30] The problem lies too deep for such superficial remedies, Genesis implies. Self-centredness is now too fundamental a characteristic of our fallen make-up.

To Adam too the curse brought a changed relationship to work, all too evident, we may remark, in our affluent industrial society. Gone, very often, is the pure joy of labour and service.[31] What the Bible is saying is not that (in the particular sphere of agriculture which was Adam's) new and nasty species of plant appeared, or that old and pleasant ones grew spines. There was rather a change of outlook and relationship (again).[32] Common and hitherto delightful tasks became suddenly burdensome, as they are always liable to when godliness and goodwill fall from their pre-eminence. Plants which man cannot turn to his own use he now regards with displeasure; they are weeds. Correspondingly man has become the exploiter of the lower animals, not their friend and benefactor.[33] The satisfying and gracious harmony between man and nature has become a discord; the ecological conflict has begun.[34] It is sin, not the Judaeo-Christian tradition, which has laid the foundation for the crisis.[35]

In all these respects relationship is the key category. As with Othello towards Desdemona an alien spirit has entered the consciousness of mankind, and pure happiness has fled. This is the condemnation.

Of course the judgement did not mean that all joy was to vanish from the partnership of husband and wife, all pleasure in work or nature to be a thing of the past, all childbearing a nightmare. God's judgement is always tempered with mercy and designed, if possible, to remedy the fault; and the sentence here is certainly far from savage, and mitigated by the immediate promise of a Deliverer.[36] Yet who can deny that there is correspondence with present actuality in the description of what is to be mankind's lot? Romantic love still holds out to men and women the prospect of sublime happiness rarely, if they are honest, ever truly realized. Work, for all its reward, is over-burdened with anxiety, rivalry, boredom or distastefulness; and so on. But these things were not designed to be so. When God made man and woman 'He *blessed* them and said . . . *have dominion*'. They were to 'reign in life'.[37] Why then are things otherwise? Why are men so often against women, children against parents, class against class, race against race, superpower against superpower? The Bible answers realistically, 'Disobedience and the Curse'.

It is sometimes suggested that the only intelligent and acceptable interpretation of the Garden of Eden narrative today is to regard it as myth, like say the tale of Prometheus. This would allow it to be 'true' (to experience) but not *historically* true. It would be the story of Everyman (as noted before), and the moral choices all men and women face in life. This cannot be regarded as a satisfying view, for several reasons. It fails to do justice to the New Testament understanding of the matter.[38] It gravely weakens the authority of the biblical testimony, an authority to which our Lord himself appealed.[39] It runs counter to the strong impression made on the ordinary reader, that the narrative is meant to be read *as history,* for it runs continuously into what is plainly so meant and betrays no suggestion that it is to be read otherwise. Most obviously, to regard it as myth (in the sense noted) is to leave biblically unanswered the cardinal question, how did our sad human situation originate in the first place? Was it always like this? Why does death get us in the end? Finally, this line of interpretation is liable to be curiously lop-sided. On the one hand scholars who hold it accept the biblical teaching that man's salvation is through a series of concrete historical events[40] ('salvation history'); on the other, the Bible's account of how humanity came to need salvation is denied historical concreteness. This is highly unsatisfactory.

At the other exegetical extreme is the view, already noted, that the curse was accompanied by great changes of *an outward and material nature*[41] in earth's flora and fauna. There are a number of strong arguments against this view. It does great violence to the language;[42] it shifts its emphasis from what matters most (the inwardness of things) to what matters far less; it compromises its credibility. I believe it misunderstands the genius of the Bible, and has a tendency to trivialize the great themes with which Scripture is dealing. Together with associated views on the magnitude and significance of the Flood[43] as a physical phenomenon it cannot be convincingly defended by sound exegesis, or, it may be added, by science or common sense. It overlooks the supreme importance of relationship.

CHAPTER NINE

Adam and Mankind

The Bible does not state in unequivocal terms that Adam and Eve were the physical ancestors of all present-day members of the human race. The relationship seems rather to have been a *representative* one, as that of Jesus Christ is. The Bible provides considerable evidence that Adam was called-out, like Abraham, to found a new race, truly human. (What this means in terms of human solidarity is left to Chapter 13).

We must come now to the question of what the Bible tells us about the genetic relationship between Adam and Eve and the human race. To be particular: does the Bible mean us to understand that all mankind has descended from this single human pair? There would be no inherent difficulty from the biological point of view in such an idea, but what we are enquiring is whether or not the Bible teaches it. It has certainly been held traditionally that it does, and it is probably true to say that this is the impression conveyed to the ordinary reader. Does the Bible make any unequivocal statements, however, to this effect?

There are, as a matter of fact, two biblical passages that might be cited. The first is the statement in Genesis 3:20 that Adam named his wife Eve 'because she was the mother of all living'; the second is Paul's assertion in his sermon in the Areopagus that God 'made from one every nation of men to live on all the face of the earth' (Acts 17:26). Before we look carefully at these there are two other (New Testament) passages which might be cited but which need not detain us long. The first is in Romans 5. Here Paul speaks of Adam as the 'one man' whose sin, as a matter of history, brought death into the world. In this Adam is contrasted with the 'one man Jesus Christ' whose obedience brings life. More pointedly in the second passage (in 1 Corinthians 15), making a similar comparison, Paul speaks of 'the first man Adam' and 'the first man' contrasting him with 'the last Adam' and 'the second man' (verses 45,47). However, the interpretation of 'first' in this latter passage depends on its being set in antithesis to 'second' and 'last'; clearly, for this reason it is not sound exegesis to read into it that Adam was the *physical ancestor* of all men. A rather similar argument applies to the passage in Romans. We return now to the passage in Genesis 3. Adam bestowed on his wife the name Eve, 'because she was the mother of all living'. These words have been taken in two ways. First, they have been regarded as an inspired comment by the narrator (traditionally regarded as Moses) contributed long after the event.[1] (In that case Adam's own motive would be open to conjecture; perhaps he was divinely prompted to choose the name[2] without knowing exactly why). If this interpretation of the clause is correct (i.e. if it is later inspired comment) then our point is probably settled, and all humanity is indeed descended from Adam and Eve. But there is a second and more natural way of taking the words. They represent what was in Adam's own mind as he named his wife 'Eve'. In this case the above conclusion is by no means so certain. The name 'Eve' is probably derived from the Hebrew *hayyā* which means 'life'. Now Adam had been warned that disobedience would mean death, and probably summary death. But when, soon after, the judgement was pronounced it referred (to his surprise?) only obliquely to death ('till you return to the ground') and spoke rather of life, though under travail, toil and hardship. Moreover, it awakened hope of a Deliverer, the 'seed of the woman'.[3] The stress, unexpectedly, was thus on life rather than death! In these circumstances what was it that moved Adam to call his wife's name 'Eve'? In his significant gesture most commentators discern faith.[4] His reaction may be vocalized, perhaps, in this way. 'It's not all up', he is saying, with vast relief; 'there's life ahead, greatly multiplied, and it will come through her!' If this is a correct understanding the stress should probably be on the *all living* rather than on the *all*. One needs to remember that Adam and Eve were still in the garden and in all likelihood had never been outside it. Moreover, they had had no children. With the outside world an unknown quantity, and procreation (except as they had witnessed it in the animals) an unknown experience it is stretching Adam's understanding rather far to insist that his motive must have comprehended worldwide humanity as we understand it. True, his parenthood of the entire subsequent human race might have been divinely revealed to him, but there is no biblical warrant for asserting this. Even the failure to find a 'helper fit for him' can firmly be related only to what was in the garden with him. So we must conclude that the clause 'because she was the mother of all living' cannot be taken as a positive affirmation that the entire human race has come physically from Adam. It is more likely to signify Adam's faith responding as centuries later Abram's was to respond, when God gave him the name Abraham, 'for I have made you the father of a multitude of nations. I will make you exceedingly fruitful.'[5] Understood thus it can be argued that it is really theologically more meaningful than as understood traditionally.

We turn now to Paul's words in the Areopagus: 'God made from one (*ex henos*) every nation (*pan ethnos*) of men to live on all the face of the earth' (Acts 17:26). In detail, the 'from one' is unspecific. The NIV supplies 'man'; the NEB and JB 'stock'. The 'every nation' could equally well be translated 'the whole race'.[6] Paul has been pressed into addressing the dilettante philosophers of Athens, and he seizes his opportunity. He knows he is not speaking to men immersed in the Jewish scriptures; rather, as a Jew, he is facing the widely different culture of intellectual Athens. So he seeks to establish as firm a base as he can for his message, a message universal for all men: judgement, repentance and faith in a risen Saviour. He faces the materialistic Epicureans (from whom he can expect little sympathy); the pantheistic Stoics (from whom he can expect a little more); and the polytheistic crowd. He starts with two points: an Almighty Creator needing no human shelter, help or provision; and the equality ('from one stock') of all men before him. He then proceeds: this God is actively concerned about men. As the Sovereign Architect of their historical circumstances he has a purpose, that all men should feel after him and find him. But clearly their seeking has gone astray; their idols, and the altar to 'An Unknown God' indicate *that*. In all this Paul is saying in a preliminary way what is essential to his central message, and nothing more; at least, it is reasonable to suppose that that is so. His quoting of two passages from Stoic poets (and none from the Old Testament) shows how anxious he is to establish what common ground he can. Now all this makes it unreasonable to insist that his reference to mankind's origin 'from one' was intended to be interpreted as 'from one human pair'. Paul was too skilled a preacher to have risked the serious attention of a less-than-serious audience by introducing such a particular and probably obnoxious red herring. He knew he was already sailing very near to the rocks of prejudice, for the Athenians boasted that they had 'sprung from the soil'. So we may conclude that whatever was Paul's own view in detail of the origin of the human race, his language on this occasion was studiously general. What he asserted, indubitably, was that all men share a common standing before God,[7] and a solidarity towards one another.[8]

We may conclude this section therefore by saying that while the biblical evidence is consistent with the view that the whole race has descended from a single human pair (indeed, that that is the *prima facie* view) it *does not positively require* that view. There are other possibilities to which we must keep our minds open.

When Cain had murdered his brother and faced the sentence of banishment and wandering he was gripped with a terrible fear: 'Whoever finds me will slay me', he said.[9] To whom did he refer with that 'whoever'? He was the eldest son, and so far as Scripture tells us, the only others of his race were his parents. It seems very unlikely that he was projecting his thoughts forward to the time when a new brother, still to be born, had grown up and conceived the idea of avenging Abel, a murdered relative he had never known. It seems far more likely that there were other beings (we will provisionally call them 'men', in inverted commas) who had never been in Eden nor sinned 'after the likeness of Adam's transgression',[10] and into whose company Cain was now compelled to move. He would naturally fear them, as all men fear the unfamiliar and unknown, especially when they have a guilty conscience.

Cain acquired a wife—from where? The traditional (and legitimate) explanation is that he married an unnamed sister.[11] But it is equally possible that he married a 'woman' from outside Eden. He had a son, Enoch, possibly his first. After this son he named a city which he built.[12] By 'city' we are not, of course, to understand a large and (by our standards) considerable centre of population; nevertheless, it hardly does justice to the word to use it of an assembly of a dozen or so primitive dwellings.[13] The overall impression of the story of Cain is then, as Kidner remarks, 'of an already populous earth'.[14] This is surprising, since the narrative does not suggest a high rate of growth in the family of Adam, but rather the reverse. As often as not sons are born only when fathers are 100 years old or more.[15]

With the opening of chapter 6 the rate of increase seems to accelerate; so does the rise of evil and violence. Associated with this (and perhaps the cause of it) is a mysterious union of the 'sons of God' with the 'daughters of men'. What the Bible means us to understand by this has long been a matter of dispute. The problem concerns the identity of the two parties involved, the 'sons of God' and the 'daughters of men'. For a good discussion of the suggestions which have been made reference may be made to Blocher.[16] The significance of the passage in the present discussion turns, however, not so much on its detailed interpretation as on the impression it gives that there were more on the earth at this time who could be classed as 'human' than simply the descendants of Adam. No doubt it is an impression only and could be construed otherwise; but it adds to the plausibility of a rather far-reaching suggestion the substance of which has often been made before.[17] Briefly, it is that Adam was the head of the human race not in the sense that all men have physically descended from him but in the sense that, before God, he *represented* them all. Let us follow this up.

In Romans 5:14 Paul makes the important assertion that Adam was a 'type' (RSV, Gk. *tupos*) of Jesus Christ, the coming One. Other modern versions have rendered *tupos* here by 'pattern', 'foreshadowing' or 'prefiguring'. Technically, Paul means that Adam, in God's design, prefigures Jesus Christ. In other words, some important aspect of our Lord's life and work is paralleled and illustrated by something of Adam's. But what aspect? Paul develops his thought in the subsequent verses of this chapter and indicates that what he is referring to is a structural built-in *solidarity,* intrinsic to the created nature of mankind, and between Adam-and-all-men on the one hand, and Christ-and-all-men on the other.[18] In other respects Adam and Jesus Christ are vastly different, but in this they are alike. Each, at some crucial point in history, stands in for the whole human race, as its representative and head, and brings it to ruin (in the one case) or to glory (in the other). To qualify for this role there are only two things constitutionally necessary, Scripture seems to imply. The first is, to be *truly human.*[19] The second is, to be *divinely appointed.*[20] Both of these things were true of Adam, and both were true of Jesus Christ. But physical ancestorship as such is not apparently a requirement in the divine economy, or Adam could not have 'typified' Christ.

Biblical analogy in other ways supports this understanding of Adam's relation to the human race. At three points in biblical history there have been striking new beginnings originating with single individuals: Adam, Abraham and Jesus Christ. Elements of a common pattern are exhibited by them all. Divinely-named and called as individuals, each was moved to a new environment to train as a pioneer—Adam to Eden, Abraham to Canaan, Jesus to Galilee.[21] Now at least in the case of Abraham and Jesus Christ, each was called to be the founder of a community not bound to himself fundamentally by ties of natural descent. In the case of our Lord this is obvious, but even in the case of Abraham the remark holds.[22] So it would not be out of character if the same was true of Adam also as we have been suggesting. It is the spiritual tie rather than the physical one which is important in Scripture, a principle illustrated again in connection with the communities of Israel, and of the church.[23]

We may summarize the conclusions of this chapter then as follows. The *prima facie* impression given by the Bible is that the entire human race has descended from the single pair, Adam and Eve. Closer inspection, however, indicates that it makes no unequivocal statement to that effect. On the contrary, it gives us very solid reasons for believing that Adam's headship of the race was a representative one, and that universal physical descent from him is not involved. It is the spiritual link which is the important one; to be 'in Adam' means something analogous to being 'in Christ'.

CHAPTER TEN

Contingency, Necessity and the Anthropic Principle

The universe seems to be a contingent structure, not a necessary one; that is, it might be other than it is. But contingent upon what— Chance or an intelligent Creator? The Anthropic Principle, a fairly recent formulation in cosmology, makes it look increasingly as if the universe was intelligently planned.

We begin this chapter with a small scenario: three boys in a classroom, one rather brighter than the other two. It is a lesson in Euclidean geometry, and the master is anything but inspiring. As he drones on, one of the boys, to relieve the boredom, draws a triangle. Next he measures the angles. Then he adds them up. The exercise is aimless, mere doodling; but it ends with something rather fascinating. So far as he can make out the sum is exactly 180°! He informs his companion. 'I don't believe you', says the other, but finding it is true, he changes his tune. 'What a fluke!', he says or—'You've done it on purpose'. The third boy had been listening. Now with a superior smile he interposes. 'Sillies,' he says, 'all triangles have angles which add up to 180°.' And with a few freehand strokes he draws a simple diagram to prove his assertion (Fig. 2).

This rather contrived scene illustrates several points. When a fact is brought to our notice we face two possibilities concerning its standing. Either it 'happens' to be such as it is, and it might have been different; or it has to be what it is from the rational nature of things, and *couldn't* have been otherwise. In the first case we call it contingent; in the second, necessary. The fact that the angles of a triangle add up to 180°[1] is, we believe, a *necessary* fact. It follows inescapably from the nature of a plane triangle and from certain propositions (the 'axioms')[2] which we hold to be self-evidently true. Its identification as a necessary fact follows from our ability to derive it by the principles of logic without any recourse to observation or measurement. A contingent fact, however, rests on a different footing. No axioms exist (we believe) from which it could be known through reason and logic alone. So far as these are concerned the actuality might have been quite otherwise. For instance, in the case of

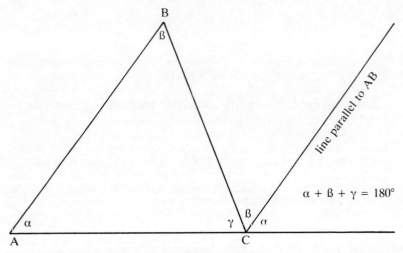

Fig. 2   The angles of all triangles add up to 180°.

our cosmos it is conceivable that the velocity of light might have been smaller than it is, and the gravitation constant larger; biological reproduction might have involved three sexes, and the genetic code might have spelt-out amino-acids very differently. Clearly, knowledge of these things has come only through empirical channels, involving actual physical observation and measurement.

Now the demonstration of contingency raises an interesting (and important) question: contingent upon what? What has selected this possibility and not that, to be the actuality? How can we account for the fact that things are what they are and not something else? One answer to this is that there *is* no accounting, it is just a 'fluke'; or (if we prefer to disguise our ignorance) that it is due to chance. The other answer is that it is *designed,* contingent upon the intention of some person; 'done on purpose' as the second schoolboy alternatively put it. To be sure, this second answer too might be a cover for our ignorance. Yet there is an important distinction between the two answers. Chance implies nothing further, beyond itself; design or purpose implies an intelligence and will, to which it would only be realistic to ascribe an immeasurable greatness. Now part of the design (if design there be) has evidently been to introduce into the cosmos a creature (ourselves) itself capable of understanding, itself a designer and purposer. For those who entertain the second answer, this raises the question (to put it no higher) whether there are any signs of possible communication from the Designer—in fact of what the theologian would call Revelation. And it is the entry of this factor into the situation that differentiates the two answers to the question 'Contingent upon what?'. Both are, as far as they go, expressions of ignorance; but there is a profound lack of symmetry between them. 'Chance' as an answer can justify itself only by appealing to ignorance; 'Design' can (and does) open the possibility of an appeal to Revelation if we can be persuaded that the Designer has given us such. And this appeal, as I have argued, can be both intelligent and consistent; Revelation is even something to be expected on the implied premises.

This discussion has proceeded on the tacit assumption that contingency of the natural world has been established. But has it? Might it not be the case that the universe is a *necessary* structure, and could not be other than it is? Several things can be said in reply to this. Of course the testimony of the Bible is wholly against it. God *chose* to create the world, it says.[3] So is the testimony of our common understanding of ourselves, for at a stroke this idea would seem to make our consciousness of personal freedom of will an illusion, and at the same time deprive us of any grounds for moral responsibility.

But independently of these very weighty considerations is another: contingency is one of the basic convictions of science. In ancient Greece when men began to speculate on the heavenly bodies it was reasoned that these bodies, being heavenly, must of necessity move in circles, since the circle is the perfect curve. This is the rationalistic, *a priori* approach. The characteristic approach of science is different. It is observational, i.e. empirical: if we want to know how the planets move we must look and see. Following this line it was discovered that, in fact, they moved in ellipses, not in circles. Later, of course, the elliptical orbits were explained as a consequence of the inverse square law of gravitation. But we have to be careful how we interpret this 'explaining'. Scientific laws, it is now recognized, are descriptive, not prescriptive; they have no causal power, they don't *make* things happen. So it would be truer to say that the inverse square law follows from the observed motions of the planets rather than to say that the motions follow from the inverse square law. And further, this 'following' is not a matter of rigid logical necessity, for as we know the inverse square law of gravitation has now been superseded (as a result of more accurate observation) by relativity theory. To give this ultimate role to observation is to recognize that the universe is contingent—i.e. not the sort of thing that follows from any final logical necessity. Confronted with this conviction then we are compelled to assign our cosmos either to 'chance, accident' or 'intelligent purpose, God'. The Bible says the latter, and it says it with great emphasis.

The Anthropic Principle

The Bible's teaching on this question (the origin of contingency) will doubtless seem to many a very reasonable one, however few allow it radically to shape their lives. It will appear all the more convincing if we give due weight to a fact which many in our scientific culture overlook or underrate: that the eager, probing, wondering, evaluating, mind of man, agitated often with guilt, fear, pain or perplexity, *is itself part of the whole to be explained*—perhaps even the most important part. For to the necessitarians man can hardly be more than an element in a complex logical edifice, while to those who assign to Chance the position of ultimate arbiter he is, strictly, just a cosmic accident, and seemingly a tragic one.[5] Neither of these viewpoints deals satisfyingly with the fact we have noted, the fact of man himself; yet this is surely the most inescapable of all facts. For the Bible, on the contrary, it lies at the very centre of its world-view.

That our universe was intelligently planned rather than that it has merely 'happened' is a view which most people, I believe, would agree has been encouraged by some remarkable findings in Cosmology. Today the favourite theory of the origin of the stellar universe is that known popularly as the Big Bang Theory, the discovery of the isotropic low-temperature background radiation having given this theory a decisive advantage over the rival Steady State Theory of Bondi, Gold and Hoyle, which has now fallen into disfavour.[6] The Big Bang Theory can be briefly described as follows. The universe started as a 'fireball' of elementary particles and radiation at an inconceivably high temperature and density; at least, this is how we would have to describe it at the incredibly short period of 10^{-43} sec. after its real 'beginning' (whatever that was). The fireball was exploding, and understandably cooling as it expanded. At first the temperature was so high that nothing but the very ultimate constituents of matter could exist; molecules, atoms and even less-than-ultimate particles like protons and neutrons (probably) would have been instantly shaken to pieces by the enormous thermal energies. As the fireball expanded and cooled the more complex structures began to appear: first protons and neutrons, then helium nuclei; and after that, at a much lower temperature, hydrogen and other atoms. Finally combinations of atoms, or molecules formed. Meanwhile, of course, on a grosser scale, the matter of the universe had been associating into galaxies and stars, some of the latter giving rise to planetary systems. There for the moment we can leave the history.

On what is this picture of things based? Firstly, on the well-established observation that the universe is expanding, probably at a diminishing rate. It is not, however, expanding *into* space; it is space itself which is expanding. The picture is fairly clear, and with the demise of the Steady-State Theory it leads naturally enough to the idea of an originally enormous density and temperature 'at the beginning'. Secondly, the organization of material systems is believed, from a vast amount of experimental work, to be subject to four basic forces. The first, *gravity,* is well known. It results in our experience of weight,[7] and is relatively very weak. Familiar too is the quite different *electromagnetic* force which holds atoms and molecules together and so gives strength to a piece of iron or wood. Then there are two other basic forces of which we have no everyday experience. They are called the weak and strong nuclear forces, and their sphere of operation is within the nucleus of the atom. Now of these four basic forces three have been satisfactorily related together, gravity alone remaining outside the unifying scheme. The knowledge of how these forces work and of the gross state of our universe at the present moment enables us to calculate—not exactly but in a very approximate way—what it was like in the past as well as what it will be like in the future, much as we can calculate when there have been or will be eclipses of the sun (only the calculation is very much less precise). This calculation can be carried backwards in time to a point at which the universe was inconceivably 'young', small, and hot. However, at present there is a limit; we cannot go back beyond about 10^{-43} sec. from the 'beginning' because of our failure to link gravity in with the other forces of nature.

Let us suppose that these calculations can be accepted, and that we may start with the universe as an incredibly hot and dense fireball, expanding with enormous rapidity. What will govern the detailed course of its expansion so far as its large-scale aspects and its physical and chemical features are concerned? Three obvious initial factors are its constitutional make-up, its total material content, and its temperature. Consider these briefly in turn. Matter exists in two forms called respectively 'matter' and 'anti-matter'.[8] They are each a sort of mirror image of the other. They can mutually annihilate each other with the production of radiation. Had these two forms been equal at the outset then it is likely that our present universe would have consisted of nothing but radiation. In fact, their initial unbalance (perhaps to the extent of nor more than 1 part in one thousand million) seems to have been of extreme significance.[9] The next two factors are related in an even more astonishing way. The course of the expansion depended enormously on the great physical constants which govern the behaviour of matter: the gravitational constant, the speed of light, the electrical charge on the proton, the mass of the electron, and about ten others. Now the staggering thing about these is that they seem to be related together in such an amazingly precise way that our universe has followed the exceedingly narrow path of physical and chemical evolution along which alone conditions could obtain which would make life physically possible. Alter one of the fundamental constants of nature and at the moment when the temperature was right for life there would be remaining practically no hydrogen, an element of supreme importance. Change another very slightly indeed and hardly a trace of carbon or oxygen would be available anywhere. Given an initial ball of matter, the energy with which it was set exploding had to be fixed with extraordinary precision if the universe was ever to be habitable:

> If the bang is too small, the cosmic material merely falls back again [gravitationally] after a brief dispersal and crunches itself to oblivion. On the other hand, if the bang is too big, the fragments get blasted completely apart at high speed, and soon become isolated, unable to clump together into galaxies. In reality, the bang that occurred was of such *exquisitely defined strength* that the outcome lies precisely on the boundary between these alternatives.[10] (my italics).

The striking thing is that this sort of conclusion crops up over and over again. Let me quote further from the same fascinating book:

> The fact that the two sides of the inequality (3.9) are such enormous numbers, and yet lie so close to one another, is truly astonishing. If gravity were *very* slightly weaker, or electromagnetism *very* slightly stronger . . . all stars would be red dwarfs. A corresponding change the other way, and they would all be blue giants . . . In either case . . . the nature of the Universe would be radically different.[11]

Or again:

> It is hard to resist the impression of something . . . possessing an overview of the entire cosmos at the instant of its creation, and manipulating all the causally disconnected parts to go bang with almost exactly the same vigour at the same time, and yet not so exactly as to preclude the small scale, slight irregularities that eventually formed the galaxies, and us.[12]

And finally:

> All this prompts the question of why, from the infinite range of possible values that nature could have selected for the fundamental constants, and from the infinite variety of initial conditions that could have characterised the primeval Universe, the actual values

and conditions conspire to produce the particular range of very special features that we observe. For clearly the Universe is a very special place: exceedingly uniform on a large scale, yet not so precisely uniform that galaxies could not form; extremely low entropy per proton and hence cool enough for chemistry to happen; almost zero cosmic repulsion and an expansion rate tuned to the energy content to unbelievable accuracy; values for the strengths of its forces that permit nuclei to exist, yet do not burn up all the cosmic hydrogen, and many more apparent accidents of fortune.[13]

We can sum up this 'catalogue of extraordinary physical coincidences' which have conspired to make the universe a place where life, as we know it, is at least physically possible by saying that it looks as if it had been painstakingly and specifically designed for man. Hoyle, whom no one would accuse of being a Christian apologist, wrote:

A commonsense interpretation of the facts suggests that a superintellect has monkeyed with physics, as well as chemistry and biology, and that there are no blind forces worth speaking about in nature.[14]

This may not be the language of reverence, but it speaks clearly enough. It expresses in a popular way what cosmologists refer to guardedly as the 'strong anthropic[15] principle.' It amounts almost to a scientific confession that God made the world for mankind to inhabit; almost to such a confession, but not quite. For there are ways of avoiding this implication, logically quite legitimate. I close this chapter with an analogy which I think illustrates fairly one such way of doing this.[16]

Two observers in total darkness happen to be looking momentarily towards a point in a high wall. Through the wall runs a small precisely-formed hole. It might be plain cylindrical, but it would make the analogy more adquate to suppose it finely threaded. In that brief moment a small object, evidently fitting the hole with exceedingly fine tolerances, passes neatly through, generating just enough gentle friction to render the event visible. 'Phew!' says the first observer; 'Someone must have made that specially and aimed it with phenomenal skill—shape, trajectory, orientation and spin dead right!' 'I don't agree', says the second. 'Didn't you notice that the wall was very slightly curved? I guess it forms part of a vast enclosure. That object has been bouncing around with an assortment of others for an infinity of time. It's only just got it right.'

Such is the nature of the choice with which the Anthropic Principle confronts us. Cold logic *cannot* settle this choice for us. In the end it will depend to a major degree on our moral and spiritual consciousness. For we, the ones making the choice, have a moral and spiritual nature, and that nature cannot be disregarded. It is indeed the most inescapable part of the Reality about which we must frame our beliefs.

It remains an amazing fact that the scientific secularist bases his whole position confessedly upon ignorance. He starts at the lower end of things, with ultimate particles. At first, as a convenient technical shorthand, he embodies his ignorance in the notion of chance. But in his programme chance soon forgets her humble origins, and begins to play God. She becomes creative, and after much labour she gives birth, finally—to the scientific secularist! So what does he think of his own consciousness and of its moral and spiritual nature? Of these things, which he knows with more immediacy than anything else, he is not quite sure. He fumbles over them, and is ill-at-ease. But at least he is sure of his ultimate particles, and sure too that ultimately, they behave quite incomprehensibly! Life, it must seem to him, is bounded at both extremes by invincible ignorance. At the lower end is the Uncertainty Principle; at the upper the Theatre of the Absurd. His creed is hardly a satisfying or reassuring one; all it can advise him to do is to join in the dance and try to enjoy it.[17] But to many in our modern society the secularist outlook is making that increasingly difficult and problematical. The Christian may be forgiven for thinking that only the most determined 'wishful unthinking' can sustain belief in such a creed, when the Bible offers a worthier one, which he (at least) finds so much more satisfying—and so much more compelling.

CHAPTER ELEVEN

Chance Again; and the Origin of Life

Secularist evolutionary thought is in crippling bondage to a mythology of chance. It has never faced the fact (illustrated by such a familiar matter as telephone numbers) that random events, confirmed as such by any standard test, may nevertheless be individually the results of fully purposive activity. Again, it has claimed that mutations are random, and natural selection the 'blind' means of exploiting them. However, even if these claims are true, *divine providence is not thereby excluded.* It is an impoverishing logical fallacy to think that a physico-chemical origin of life, Darwinian evolution, man's descent from 'apes', and even a completely naturalistic cosmogony banish the possibility that God is in firm control.

But *are* mutations random? No serious evidence has ever been given for this key secularist presupposition, and it is difficult to see how it could be. It forms the basis of a merely methodological principle—one that may prove to be leading us in the wrong direction.

In a previous chapter we discussed the biblical attitude to chance, and we saw that the Bible recognizes and indeed uses the category, but sees in it no obstacle to its teaching on the sovereign providence of God. God does not play dice; he directs them.[1] Yet their fall may with entire legitimacy be regarded by the human observer as random, the result of chance. Lest this seem an impossible position to maintain let me justify it by a simple illustration.[2] The final digits in a column of numbers from the telephone directory constitute a sequence which would probably[3] stand up to any statistical test for randomness which we cared to apply. We may suppose that the numbers were allocated by a Telephone Authority on principles of its own choosing based on a detailed knowledge of its customers. We have here then a system in which statistical randomness results directly from deliberate, intelligent, purposeful activity on the part of a superintending Authority. If it be asked whether a powerful enough computer given access to the same customer-information as the Authority could not probably extract the principles on which the

latter worked (and so could overcome the element of unpredictability
in the numbers and their claim to randomness) the answer is, 'Yes,
quite possibly'; but this does nothing to negate the point made by the
analogy. For God has knowledge inaccessible to men, and his wisdom
is inscrutable to men. The objection just noted, based essentially on
the finiteness[4] of the human Authority, therefore loses its force when
directed at Providence. The biblical position is quite logical. There is
nothing incoherent in the belief that random numbers or chance
events, though properly so called, are nevertheless individually willed
by God.

Now chance, or randomness, plays a great part in the currently-
orthodox theory of evolution. It is especially prominent in the
supposed origin of the small inherited variations which are regarded
as the raw material for natural selection.[5] I shall later give some
examples of the emphasis that leading evolutionary writers place on
the randomness of these variations; but at the moment I wish to make
two comments. Both the secularist and the 'creationist'[6] commonly
regard randomness (or chance) in this context as the emphatic
negation of purposive divine activity. If natural selection has
operated on random variations then this is supposed to exclude God.
He can have had nothing to do with the marvellous diversification of
living things in the course of earth history. That is their conclusion.
But it is a false one, as I have just attempted to show. For if a
sequence of digits from the telephone directory, confirmed as random
by any available test, can, nevertheless, be the result of careful,
deliberate and intelligent choice by a superintending Authority, then
their argument is seen to be fallacious. Even if gene mutation, gene
recombination, and chromosome changes are truly random,[7] divine
providence is not thereby excluded from the evolutionary process.
This needs to be clearly understood. Further, the 'creationist' needs to
be reminded that the Bible itself strongly supports this conclusion.
He need only recall the random arrow that slew king Ahab,[8] or think
of Proverbs 16:33 to be reminded of this.

My second comment is on a matter of great significance. The
orthodox claim is that gene and other inherited variations are
random. By this is meant that they have no relationship to the
direction of evolution. Not all evolutionists believe that this is always
so; for instance, some believe there is incontrovertible evidence for
what is called orthogenesis, or straight-line evolution, a sort of
momentum carrying evolution forward in an already established
direction and independently of natural selection.[9] However, that
some evolutionists believe this need not detain us; we are concerned
to examine the paramount claim that such variations are random. An
enormous superstructure of evolutionary philosophy has been

erected on this claim, and it is of extreme importance to the secularist's case.[10] Can it be substantiated? It should be apparent at once that if the claim could be shown to be false—if it could be demonstrated that the direction of variation might itself be related to the end in prospect—then the whole outlook of theoretical biology would have to change dramatically.[11] It is not too much to say that a revolution would be involved of magnitude comparable to that which dethroned classical physics. In these circumstances it is time that those who believe in the randomness of variations should give some thought to providing evidence for their belief. But before we go on to discuss the matter further let me indicate how I believe we should regard this particular doctrine.

Consider a party of men finding themselves in a tract of unknown country with no familiar landmarks. Their problem is, to reach civilization. But how to proceed? They look around and choose what appears to be the easiest and most promising direction, say due north; and they set out. Wisdom clearly dictates that having decided on this direction they should stick to it, lest they wander in circles; but it equally dictates that they should be alert for unexpected clues, like traces of previous wanderers or even the persistent hunches of one of their number. But, of course, they will not easily be diverted from their initial plan—to travel north. It constitutes for them a *methodological principle,* not to be lightly set aside. However, they must recognize that it may prove in the end to have been the worst possible choice. Such is the rationale, I believe, of regarding variations as random. It is a methodological principle (like the one just illustrated); no more. It is not an established truth, nor even a working hypothesis. Like the decision to go north, it may be leading us away from our goal, not towards it.

Now we must return to our discussion of chance and randomness. Let me begin with a sentence of Jacques Monod:[12] 'Pure chance, absolutely free but blind, [is] at the very root of the stupendous edifice of evolution.' Of course, Monod is erecting a mythology; Chance for the ordinary reader (if not for Monod himself) is here half-way to becoming a blind goddess living unfettered in a basement from where she issues her directives. She is, he goes on to imply, the 'creator of *absolute* newness'.[13] She has a 'realm', from the 'products' of which she supplies 'nourishment' for natural selection.[14] She appears to run a 'vast lottery, in which natural selection has blindly picked the rare winners from among numbers drawn at utter random'.[15] It all sounds rather like a Babylonian creation myth though with respectable scientific overtones. But consider for a moment: what could Monod have meant by 'Pure chance, absolutely free'?

Suppose we have a die, with faces numbered one to six as usual, but in the form of a light hollow cube with a heavy pellet suspended at its centre. This is thrown repeatedly by a mechanical device. The number six comes up a fraction of the total trials which moves progressively nearer to ⅙ as the experiment proceeds. 'Pure chance, absolutely free'? Presumably. But now let the pellet in the die be moved nearer to one face. The experiment proceeds, but the proportion of sixes becomes quite different. Still 'pure chance, absolutely free'? Monod would have had to say, 'Yes', for the case is not essentially changed.[16] But someone else seems to be cramping her style! What valid meaning can now be given to the epithets 'pure', 'absolutely free'? It looks as if chance has to take what she is given, and the results she turns out can be made to be anything, statistically speaking, that the experimenter wishes. He has only to build-in suitable features to the system she works with, and his desired goal is achieved. Chance appears to be, shall we say, not a sovereign goddess but almost a captive robot, with someone else in charge above.

However, it can be argued from this illustration that the experimenter has control only over the *statistical* outcome; he can't control the outcome of an *individual* throw. And since a single throw may lose or win a fortune, doesn't this leave chance with an influential enough role? This is indeed a valid point, but not a serious one. It uncovers, in fact, a curious situation. It is this. We can establish that chance is a valid concept in a given context *only by doing an experiment statistically.*[17] It is jumping the gun to conclude that this gives us the right to assume that in individual instances 'chance' is *still* the appropriate label, in the sense that individual outcomes cannot be the result of intelligent purpose.[18] No one on the basis of regularities in the statistics of burglary would maintain that it was right to assume that particular cases were not deliberately and carefully planned. As a foundation for the revolutionary and iconoclastic philosophy of life Monod is urging, one surely needs something firmer than this. Its logic is all adrift.

It is, in fact, quite impossible ever to be sure that a particular happening is the result of what Monod calls 'pure chance', and this can be argued from the very example that Monod presents to establish the contrary. He relies on what he calls 'absolute coincidences' arising from the 'complete independence' of two chains of events whose convergence produces an accident. He uses this anecdote.[19] 'Dr. Brown sets out on an emergency call to a new patient. In the meantime Jones the carpenter has started work on repairs to the roof of a nearby building. As Dr. Brown walks past the building, Jones inadvertently drops his hammer, whose (deterministic) trajectory happens to intercept that of the physician, who dies of a fractured skull. We say he was a victim of chance. What other term fits such an event, by its very nature unforeseeable? Chance is obviously the essential factor here . . .'. However, in spite of the careful design of this anecdote, Monod quite fails to prove his point. Let us ask a simple question. *How do we know* that the two chains of events have 'complete independence'? Merely to assert that they have is to beg the whole question! Give the anecdote to any competent novelist and he could spin a dozen stories round it establishing quite the contrary. Unless we know that all these stories (and an infinite number of other untold ones) are false we cannot possess this vital information. Chance remains a legitimate concept therefore, only so long as our ignorance holds; so long, that is, as we know of no connection between the movements of Dr. Brown and Mr. Jones. We are back to square one.

There is a further important characteristic of the scientific use of this concept. How do we, in fact, establish statistically that chance *is* a valid concept in a given situation? The procedure always entails the identification of what we call *a priori* probabilities. If expectations based on these turn up in our experimental results to a high enough degree of accuracy then we conclude that 'chance' is the valid concept. But this conclusion depends on whether we have identified the *a priori* probabilities correctly, and this is a rather subtle point. There are two difficulties here. The first is a theoretical one, the second practical. Let me illustrate the first like this. In the experimental throw of a perfectly uniform die the *a priori* probabilities that each of the six faces should be uppermost are equal. We can argue this from the symmetry of the system. But there is an important proviso: ideally the faces must be indistinguishable.[20] Why? Because if they are physically[21] distinguishable the possibility cannot be excluded that lurking in the wings as we do the experiment is some physical detector capable of triggering a physical impulse if the die threatens to fall with a particular orientation. Unless we know positively that there is no physical agency of this sort within striking distance our whole programme is subject to a degree of uncertainty. But we *cannot* know this positively;[22] all we can say is that we have carefully considered everything we can think of and we don't know of one. It thus appears again as if the claims of chance are ultimately dependent on human ignorance. Nor can it be said that this uncertainty is a matter of such peripheral importance that no scientist would ever be bothered by it. Give him the task of deciding experimentally whether a man who claimed that he could influence the die by will-power[23] was deceived or not and the possibility of unrecognized physical agencies might become a very live issue.

The second difficulty is this. It is often very hard to establish what are the *a priori* probabilities. The case of a biassed die is hard enough. When a gene mutates, in what directions can it go? and what are their *a priori* probabilities? Unless we can specify them we have no means of confirming that the mutation is random. The evolutionist, to be sure, uses the term 'random' in a rather special sense,[24] but he still brings in the notion of chance with all its associations. On either the physicist's definition or the evolutionist's, the claim that gene mutations[25] are random (or for that matter, non-random) seems equally impossible of substantiation. To build a whole philosophy of life on the truth of this claim is to build indeed on a quagmire. It is time the secularist faced up to this.

The spontaneous generation of life

Orthodox evolutionary theory carried to its ultimate conclusion proposes that life originated by spontaneous physical and chemical processes from inorganic materials. One form of the theory suggests that the earth's atmosphere was once what a chemist calls 'reducing', with no free oxygen present, but rather ammonia, water vapour, methane and hydrogen. Atmospheric electrical discharges produced aminoacids and other organic compounds. These accumulated as a thin soup in shallow seas. Gradually organic macromolecules developed and eventually rudimentary cells. Life had appeared.

This theory faces enormous difficulties, of which the origin of the genetic code is merely one. However, I do not wish to discuss these but rather to consider the whole matter from another angle. Let us suppose then that the theory is true. Further, let us suppose that the whole drama of the appearance of life has been video-recorded by visitors from space, so technologically advanced that they have been able to record events not only on the macroscopic and microscopic levels, but by means of sophisticated electron microscopy on the molecular level as well. The recordings have somehow fallen into human hands, and they can be played back at rapid speed so that the whole prehistory of life's appearance can be seen at a sitting. We imagine a 'creationist' and a secular evolutionist sitting down to view them together.

As the show proceeds it appears that aminoacids have indeed been produced by electrical discharges, and that later they have joined up to form first polypeptides and then proteins. The other key organic molecules, such as the sugars and nitrogen bases are also seen to have made their appearance, by what look like natural processes. 'Heteropolar' molecules became organized into films and membranes; 'myelin figures' and 'coacervates' then apparently formed; and, with a suddenness that depends more on their powers of recognition than on the recorded sequences, the viewers see what is clearly a primitive cell. The recording ends.

The evolutionist turns to his companion with unaffected sympathy; he can see he is very disappointed. 'Well, friend,' he says; 'if anything proves my case that does. Like Laplace, we have no need of God to explain things'. The 'creationist' does not reply. He is profoundly shaken, but like all wise men he prefers to think things over before he makes up his mind. It's a good thing he does so, because this sort of debate has been bedevilled from the start by hasty judgements, such as, for instance, the one of which I have made the evolutionist guilty. Let me explain.

My secular evolutionist sat down to the viewing with a definite idea in the back of his mind of what things would look like if God were really at work. Perhaps a Figure, full of numinous quality, would appear and manipulate matter visibly? Perhaps some macromolecules would break into sudden motion, clearly unnatural, and arrange themselves into a new structure, like members of a performing troupe pushing purposefully through a Bank-holiday crowd to assemble on stage? If neither of these suggestions is acceptable to my secularist reader let him formulate in his own mind what *he* would have expected to see if God had indeed been at work. For unless the secularist is able to say what he would have feared to see he hardly has much cause to congratulate himself if he doesn't see it.[26] Perhaps both he and the 'creationist' expected the recording to show at one instant a dilute soup of organic molecules, at the very next a perfectly-formed recognizable cell, living and active? Each man must answer for himself; but I have the strong impression that if he does so it will reveal that the trouble is that both are working with a non-biblical conception of God. It is not the God of the Bible of whom they are thinking. They are thinking rather of a God who plays no part in ordinary continuing physical events; there is no need for him to do so. Nature has been well-designed and constructed and proceeds satisfactorily on fixed built-in principles! Only when he desires to introduce some novelty, they suppose, need God concern himself, and when he does this his interference is at once apparent as a physical discontinuity.[27] The God of whom Scripture speaks is quite other. He is the Author, and the continuing Author, of our on-going universe and its life. Scientific laws represent only our groping efforts to trace out the pattern of his majestic ways. He does indeed, and for good purpose, sometimes work outside his usual pattern in the mode we speak of as miracle; but he is just as much present in the ordinary and the usual as in the extraordinary and the unusual; the only difference, as we noted before, is that he isn't so obviously and significantly present.[28] Once this is grasped it becomes a positive pleasure for the believer to view my hypothesized video-recording time and again. It only drives him to more heartfelt worship.[29] As for the secularist, viewing it is as much likely to convert him to a new way of thinking as to confirm him in his old way; in fact, I think, more so.

I cannot forbear in closing this chapter from asking again a question I have already asked, and repeating a comment I have already made. All secular evolutionists with whom I am acquainted make the assertion more or less positively that mutations are random, that is, unrelated to the needs or destiny of the species. *How do they know?* What evidence can they produce in support of this claim, *quite fundamental to their whole position?*[30] Can they even outline an operational procedure which might, in principle, establish it? If not, they are surely blind leaders of the blind. To *treat* mutations as random is, I maintain, a methodological principle which is natural and appropriate in the circumstances. I would not quarrel with its tentative use as such within Science. But it must be clearly recognized that as the basis for a philosophy of life it is entirely possible that the assumption is false and has all along been leading us in the wrong direction, away from a real understanding of the mystery of human existence and not towards it. *This cannot be emphasized too strongly,* especially in the teaching of the subject, whether to schoolchildren or to students.

But further: even if mutations should prove (by any objective test open to us) to be genuinely random (in the biologist's technical sense), *this would still not invalidate* the claim that they were providentially ordered. The analogy of the telephone numbers establishes this.[31] It remains therefore that if creation and providence are matters of faith, secular evolutionism with its exaltation of chance is no less so—or better, is a matter of 'unfaith'. We must choose between them on the basis of criteria quite other and wider than those which Science can provide.

CHAPTER TWELVE

Darwinism To-day

Darwin convinced the scientific world that 'descent with modification' (i.e. evolution) was (and no doubt still is) a fact. This conviction has remained steady and almost universal; but his proposal about the mechanism involved (natural selection) has been through great vicissitudes. Early in this century its acceptability fell to a low level, mainly owing to the influence of Mendelian genetics. Then in the 1930's a reconciliation took place and the Synthetic Theory (Neo-Darwinism) became established as the scientific orthodoxy. It was (and is) still based on natural selection. However, immense difficulties remain, stressed particularly by non-specialists. The biblical doctrine does not stand or fall with the fortunes of the theory.

The plant and animal worlds (especially the latter) exhibit a fantastic amount of variety. Organisms of all shapes, sizes, colours, textures, habits and life-histories throng the meadows, forests, deserts, mountains, rivers, lakes, caves, shorelines, oceans and aerial spaces of our marvellous world, and throng them in a richness impossible fully to comprehend. Where did all this diversity come from? It is a natural question, and from quite early days there have been two answers. On the analogy of things like mineral substances and metals and precious stones the diverse organisms, it has been thought, have always shown the specific distinctnesses that they now show. This is the view commonly called Special Creation. On the other hand, on the analogy of human life, where children can differ considerably from their parents and where men and women from foreign climes differ even more, it has been thought that in the course of time organisms have diversified in a continuous fashion as one generation succeeded another. This second answer is favoured by the fact that it is based on a closer analogy than the first; it compares living things with living rather than with non-living. But for centuries it lacked popular appeal because within the time available for personal observation—the mere threescore years and ten—species don't seem to change. Cats remain distinctly cats, and mice mice. It became much more plausible and attractive when the view gained ground that the earth had a much longer history than had hitherto been thought. Then indeed the stage was set for it to make rapid progress. Today it is without doubt the majority view. The enormous variety of living things is due to what Darwin called 'descent with modification'. It is this process of diversification, generation by generation, that is denoted by the word 'evolution'. At least, that is so in the English speaking world.

Evolution in this sense is by no means a new idea. It can be traced back in a very rudimentary form even to the Greeks, who seem to have thought of most things. Much later we find men such as Robert Hooke the microscopist (1635–1703), Goethe the poet and nature-philosopher (1749–1832), and Buffon the encyclopædist of natural history (1707–1778), toying with it. Darwin's grandfather Erasmus (1731–1802) went further, but suggested no mechanism. It was the 'greatest systematist of his age', the Frenchman Jean Baptiste Lamarck (1744–1829), who first put forward a detailed theory of how evolution might have taken place. His theory, now usually referred to as the 'inheritance of acquired characters', or of the 'effects of use and disuse', still, in fact, attracts spasmodic support; indeed Darwin himself invoked it for cases his own theory seemed unable to accommodate. But for the most part it has been discarded. Neverthless, as the first serious theory of evolutionary mechanism it is worth pausing for a moment to look at it. It proceeds on the following lines.

Consider a blacksmith. He develops his muscles by hard work. Part of the self-improvement (the theory supposes) is passed on by physical inheritance to any sons born to him subsequently. If the trade is followed in turn by these sons (and so on) for a sufficient number of generations the process will be cumulative, and in time a breed of men stronger-armed than usual will result. It is as simple as that.

The trouble with this attractive and easily-comprehended theory is twofold. First, in spite of intensive search there is no well-authenticated evidence that characteristics acquired in this way are ever physically transmitted, and there is a great deal of evidence that they are not.[1] Second, knowledge of the origin, minute structure, biochemistry and biophysics of the germ cells reveals no plausible way in which they might be influenced to pass the characteristics on. True, these objections may not be absolutely final; but today they weigh very heavily indeed with the great majority of biologists. Even in his own day Lamarck's theory gained little support. Its status was purely speculative.

Very different was Darwin's theory, which came a half-century later. Darwin set enormous store on evidence, and held speculation on a tight reign. His theory thus rested on an altogether more secure basis than Lamarck's. It can be illustrated as follows. Among the blacksmith's sons (numerous we may suppose, to give point to the illustration) some will be luckier in having inherited muscles better than the family average. Others, less fortunate, will have muscles worse than the average. If the sons (most or all, again to give point to the illustration) take up smithing, the stronger ones will do better than the weaker, and with larger incomes (we may suppose) will be able to raise larger families themselves. Thus over the course of many generations repetition of this pattern will automatically ensure that the blacksmith community became more muscular. The result is the same as that provided for by Lamarck's theory, but the mechanism is entirely different. Of course, this illustration is rather contrived, but it illustrates the mechanism well enough.

Darwin's theory is thus, in essence, extremely simple; it is quite within the powers of the layman to understand. It starts from two common observations; that offspring vary in unpredictable ways from their parents; and that normally each generation produces too many offspring (often vastly too many) to survive to maturity and themselves reproduce. The conclusion therefore is that between the offspring there is, metaphorically, a struggle for existence. If among the variations in the offspring there are favourable ones, then the lucky individuals with them will tend to leave more progeny than the unlucky ones without them; and if these favourable variations are heritable (as is undoubtedly often the case) then the whole population will gradually change, generation by generation, in the favourable direction. Those are the essentials of the matter.

Two points should be noted especially. First, the variations in the offspring were regarded by Darwin as random, unrelated that is to any pre-determined or consciously-sought direction of evolutionary change. Second, he thought of the evolutionary movement as directed automatically towards a single goal, success in engendering offspring. Together, these two points meant that any characteristic of a living organism (lungs to breathe air, warm bloodedness, even eyes to see) which can plausibly be regarded as conferring on its possessor an advantage in the matter of raising a large family can *ipso facto* be explained by the working of an unthinking principle ('natural selection' as it is called) operating on variations attributable to chance alone. This devastating conclusion worried Darwin himself, and no wonder. It apparently leads inevitably to the inference that life is fundamentally meaningless. Man must struggle as best he can to give it meaning. But in Darwin's day this verdict was still over the general horizon, and so was not troublesome overmuch.

This is not the place to answer the glib and shallow *philosophy* of 'Evolutionism' to which the undoubted strength of Darwin's biological arguments has lent support; that is done elsewhere in this essay. Here we are discussing Darwin's theory itself. Darwin was a superb naturalist; highly observant, painstaking, industrious, single-minded, honest, cautious, with a vast capacity for absorbing and weighing his facts. The passage of time has served only to confirm his reputation, and his lasting fame is assured and well-deserved. He knew the scientific weaknesses of his theory, and was prepared to face them. One of the principal ones turned on the question of the origin and heritability of the all-important variations; and as this has had a profound effect on the subsequent vicissitudes of his theory we will briefly consider it.

Imagine a population of glass bottles, each partly filled with ink of a different depth of colour; this difference constitutes the variation. We need not bother to suppose the bottles to multiply, but we imagine them to 'breed' by the contents of a pair of bottles being mixed and then re distributed between the two. This happens with the bottles being paired at random throughout the population. It continues again and again with them changing partners each time. Clearly, if this process is continued long enough, the original state of affairs will give place to one in which all colour variation has vanished. There will remain no bottles containing pale water, nor others with dark ink; all will show the average hue. This model, in fact, illustrates what is called 'blending inheritance', and it represents the view which was held in Darwin's day of the manner in which biological traits are passed on from parent to offspring. Clearly, it means that variability is always tending to disappear, and fairly rapidly too. It constituted a great problem for Darwin, for whose theory, as we saw, variability is a prime necessity. He never solved it. Ironically it was solved by the work of Gregor Mendel, a contemporary of Darwin, whose published results were unnoticed till 1900, eighteen years after Darwin's death. Ironically too, the science of Genetics to which Mendel's work gave rise, for nearly a quarter of a century threw great doubt on Darwin's theory, which in consequence suffered a partial eclipse. However, we must not mislead. Loss of faith in Darwin's theory did not mean loss of faith in evolution. *That was never the case*; confounding of the two is a frequent mistake. Since Darwin convinced the world of science that evolution had really taken place the idea of evolution has always been in the ascendant. It still is, however much biologists (and other men of science) waver in their convictions as to how it has been brought about.

Mendel's theory, as it impinged on the question of variation between offspring, can be illustrated as follows. Consider a characteristic like tallness or dwarfness in peas, under the control of only a very few (say one or two) of Mendel's genetic factors (or *genes* as they are now called).[2] An individual pea plant can be represented by a bottle (as before) containing not a liquid, but a very few colourless marbles, which represent the genes in question. We imagine a collection of such bottles standing for a population of pea plants. Before we set them 'breeding' we replace one of the colourless marbles here and there with a black marble. This represents a variant form of the gene; perhaps it has arisen as a 'mutation', an accidental change following a chemical or physical disturbance. We will suppose it promotes tallness. Thus the bottles in which we have placed a black marble become, symbolically, tall pea plants; the others remain dwarfs. We now set the bottles 'breeding' as before. However long they go on it is clear that uniformity will never result as in the previous case. The same amount of variability, more or less, will persist for ever. This is a consequence of the fact that the character we are thinking about (i.e. height) is represented by the sum total of just a *very few* discreet elements (the marbles) which are distributed unchanged as wholes. This rules out, naturally, any suggestion of 'blending inheritance'; the variability can never disappear. Darwin, had he known of it, would have shouted 'Hurrah!'; but alas, while his theory gained in one way it lost very seriously in another. Why this was so we must briefly consider.

Darwin knew of two sources of variation: continuous variation of unknown origin (continuous because found in all degrees from the maximum down to zero); and discontinuous variation. The latter was known in the form of big jumps, or saltations, as when a bush rose suddenly 'sports' a climbing stem, or a new breed of sheep with short legs suddenly originates. Darwin discounted these sudden discontinuous leaps, and pinned his faith instead on the small continuous variations. All his meticulous observations and years of patient study had indicated that natural selection worked with these rather than the others. But Mendelian theory taught that all inheritable characteristics were passed on, in fact, in a discontinuous fashion, like the marbles; and this seemed to mean that variation must be essentially discontinuous, often very discontinuous.[3] The establishment of Mendelian genetics on a firm experimental basis (which Darwin's theory could hardly aspire to) therefore led to a long period in which Darwin's ideas were in partial eclipse, and many great names can be cited against them. *Evolution they still believed in,* but not Darwinism.

It was not until the second quarter of this century that the apparent antipathy between Darwinism and Mendelian genetics began to be resolved. The realization that made this possible was that genes worked together in large numbers; it was the whole 'gene complex' that must be considered, not individual genes in isolation. To take a simple case: if a character (like size) is under the control not of one or two genes (as we supposed in our model) but of a hundred, then clearly variation could be much more continuous. One gene might be in the form promoting size, or two or three up to the full hundred, with correspondingly greater effect. However, things are in general much more complex even than this. The organism is rather like a social gathering, where the behaviour of any one guest may be markedly dependent on who the others are. On one occasion he may have a lot to say for himself; on another he may be much more subdued. This realization, as has been remarked, began to transform the impact of Mendelian genetics on Darwin's theory; and somewhere in the 1930's a new Synthetic Theory of Evolution, often called Neo-Darwinism, began to find general acceptance. The outstanding mathematical work of the 'population geneticists', Fisher, Haldane and Sewall Wright had a great deal to do with this, and today Neo-Darwinism is still the prevailing scientific orthodoxy. What Neo-Darwinism supposes is that the raw material on which natural selection operates is provided by inherited variations due to sudden discontinuous mutations, or to rearrangements of the genetic material, all *complexly interacting* with the whole of the genetic complex. (Sometimes, but not always, the result simulates Darwin's continuous variation). The synthetic theory is therefore still justly known by Darwin's name. Darwin's key idea and Mendel's, it is believed, have proved not conflicting, but complementary.

Does this mean than that the wider scientific fraternity is now satisfied that the answer has been found to the riddle of how evolution has occurred? Before I attempt to answer this question I must disclaim any specialist knowledge. I am a biologist, but I cannot speak as an expert. However, I will attempt an assessment. It is probably true to say that those whose work is most intimately concerned with the subject—the population geneticists, the evolutionary systematists and so on—do feel satisfied, though they would insist that we have a long way still to go. Other thoughtful biologists and those from disciplines further afield may not be so sure.[4] Many readers will know, for instance, of the sarcastic comments of the cosmologist Sir Fred Hoyle, who uses arguments based on probability but of questionable validity (as indeed most of such arguments are).[5]

This unsatisfactory position arises like this. Evolution concerns a unique *historical* process; consequently evolutionary theory (as it concerns the past) can never be tested in the way in which such theories as Relativity or Quantum Mechanics are. At the end of the day always and inevitably we find ourselves up against the question, 'Do I find this credible?' Consider as an illustration the well-known case of the vertebrate eye. Can it really be believed that such an amazing organ, with so many complexly interacting structures and functions integrated together, has resulted from nothing more than chance variations acted on by the unthinking process of natural selection? Darwin himself, as is well known, stumbled at this suggestion, though he came down finally with the answer, 'Yes.' But such a suggestion can never be proved. It will remain to the end of time either credible or incredible.[6] And it is pertinent to remark that what one finds oneself able here to believe (or forced to disbelieve) is dependent not so much on the scientific evidence as on one's basic presuppositions. Are scientific categories ultimate, or are there categories behind and beyond them? Is there or is there not, a Creator? And if there is, how far can naturalism (that is, the methodology of science) be pursued before one comes up against a full stop? These are not insignificant questions though they are often treated as such by the over-confident secularist.

The reader who is interested in the difficulties faced by the new Synthetic Theory can hardly do better than consult *The Great Evolution Mystery* by G. Rattray Taylor.[7] The author writes as a convinced secularist. 'If there is one solitary fact which emerges distinctly from evolutionary studies it is that evolution is not the execution of a consummate overall plan, divine or otherwise', he says, with supreme self-assurance. 'There have been far too many false starts, boss shots and changes of intention for that . . . many forms . . . —such as man—turn out on examination to be very poorly designed.' This of course is just foolish claptrap, though its style is fairly common in a certain class of secularist writer; but the book does also systematically and ably catalogue the serious problems facing the current Neo-Darwinian orthodoxy.

Of the problems which can be appreciated most easily, that of missing links is probably one of the foremost. Let us pause to consider it. Suppose a capable artist has produced a moving film illustrating evolutionary change. Before our eyes he has serialized the transformation of a fish into a reptile, in a manner consistent with orthodox theory. Every frame in the film sequence purports to represent a stage through which the living stream of vertebrate life has actually passed. Why then can't we find a whole range of fossils which, arranged in order, answers satisfactorily to the whole sequence or to one like it? This is the problem of the missing links. Where are they all? There are cases where the fossil record does yield many individuals exemplifying continuous change (such as the oysters), and there are cases where two widely separated groups appear to be connected by a more-or-less isolated intermediate form (such as the reptiles, birds and *Archæopteryx*); but these are not very substantial offerings in the face of what the theory seems to demand. The difficulty is by no means a slight one, however cursorily the secularist may brush it aside. It provoked the eminent palæontologist and geneticist Richard Goldschmidt to propose his theory of 'hopeful monsters'[8]—that many evolutionary changes (such as the acquisition of hair in mammals, or feathers in birds) took place not in gradual stages by natural selection, but in one fell swoop, by a sort of gigantic genetic leap. More recently has come the theory of 'punctuated equilibria' associated with the names of Niles Eldredge and Stephen J. Gould. This supposes that a very common occurrence in evolution has been a long period of relative stability (in which a form leaves plenty of fossils) followed by a brief period of considerable and rapid change (leaving very few), and followed again by a period of stability. This would account for oft-repeated observations akin to that which first attracted Eldredge's attention: a form of trilobite (a sort of giant woodlouse) abundant in a fossil bed suddenly dies out and its place is taken by a different form with no indication of any transitional forms between. Causes for such periods of very rapid change have been suggested (for example, the isolation of small populations, or some great catastrophe); but the inevitable credibility question for a historical theory remains; nothing in the past can be proved. As an explanation of the overall problem of the lack of transitional forms (especially between major groups of animals and plants) 'punctuated equilibria' will appear to many but a feeble suggestion. It is, in fact, being hotly contested.

At the base of the edifice of orthodox theory lies the intractable problem of the origin of the genetic mechanism itself, and with this our discussion of difficulties must end.[9] The problem is quite plain. In its simplest form it is this: the DNA 'tape' requires enzymatic proteins to fabricate it. But until it is fabricated the proteins cannot be made! And that omits the whole question of establishing the code, from scratch; and working out the astonishing mechanism for its near-impeccable operation. No matter what explanation is put forward, we stumble against the problem of credibility again. At this level can we really believe the explanation of naturalism?—*really*? or is it more reasonable to invoke more directly a Reality beyond nature?

It would not be fair to leave this account of 'Darwinism today' without some comments on the other side. The fossil record, we must agree, does bear irrefutable witness to the fact that a vast range of animal and plant species have existed on the earth of which no living members remain today. Indeed, so great is the variety of fossil forms now known that it is estimated that extinct species outnumber living ones by at least ten to one.[10] But the record clearly does not indicate an incredibly rich variety at the start, of which only one tenth now remains. What it does indicate is that as some forms died out others appeared. The theory of evolution makes sense of this at once; the idea of 'special creation' (as earlier defined) doesn't—at least, not without biblically-unwarranted sophistication.

Another very striking fact about the living world is its unity. This is true in at least two ways. Biochemically, a vast range of molecular groupings and enzymatic systems turns up repeatedly;[11] genetically, the significance of DNA and the genetic code are universal. No doubt a case could be made out for these things being highly desirable (and so explicable) on the basis of 'special creation'; but I must confess they seem to me more naturally meaningful and understandable if living things are related in the evolutionary sense. Perhaps we could add a third way in which the living world is a unity: systematically. For living things can to a large extent be classified in a hierarchical fashion: species, genera, families, orders and so on. Isn't it an entirely natural impression that all species of rose say, or of iris, must be *related,* and not just in the sense that their flowers have the same form, but in the sense (being living things) that their genealogies must join up somewhere as we trace them back? And if genera thus comprise species from a common origin, why shouldn't the same argument hold for families, orders, . . . and where does one stop? And that of course means evolution.

It is time this chapter concluded. It has attempted to give the reader a fair resumé of the present standing of the currently orthodox theory of evolution, Neo-Darwinism. It has possibly left him perplexed, even ill-at-ease. Evolution, at one moment it may seem to say, is utterly indisputable. It stares you in the face. But the next moment there are doubts. Why don't the great animal and plant phyla join up in the fossil record? The gaps are enormous. We find just enough significant links (such as *Archæopteryx*) to hold out promise, but nowhere near enough to satisfy. It is all devilishly teasing! Darwin once remarked that the origin of the angiosperms (the flowering plants) was 'an intolerable mystery'. It still is, as are many other comparable origins; and that after well over a century of intensive search and research. Then as to mechanism. There seems no acceptable alternative to the synthetic theory. But while the specialists regard it as entirely competent to cope, many highly qualified non-specialists have the gravest of doubts. At the lower end of the evolutionary ladder, where the debate concerns the origin of life from non-living matter, there is no consensus. Moreover, the determination to push naturalism to the extreme limit is so great (here as in cosmology) that no end to the controversy with secularism can be foreseen through the latter's sheer exhaustion. However, I must remind my reader that the objective of this essay has not been to dislodge Darwinism, nor even the dogma of the spontaneous generation of life. It has been to commend the biblical doctrine of creation. That doctrine, as I have attempted to show, does not stand or fall with the fortunes of evolutionary theory. It stands on the authority of the Bible, and of its own superlative merits. The vicissitudes and uncertainties of evolutionary theory therefore need not detain us. They only highlight the Bible's claim to mould our thinking, and make it more urgent and unchallengeable.

CHAPTER THIRTEEN

Taking Stock

The threads of the biblical and scientific evidence are drawn together and an attempt is made to relate them. The biblical picture of Adam and Eve and their sin, and of death consequently passing to all humankind, is not incompatible with current *scientific* (as opposed to philosophical) thought. Further, what has been inferred from biblical teaching may point a way out of some theoretical difficulties, even if it be at the cost of acknowledging that the methodology of science has, at this point, reached a dead end.

It is now time for us to begin to draw in the threads of our argument in the light of the objective stated at the beginning, namely to commend the biblical doctrine of creation in an age dominated by scientific assurance and theological diffidence. To what extent have we been able to do this? Let us approach the question by recalling some of the elements of the biblical teaching which relate to our dispute with the world of secular thought. They may not in themselves be the most important ones, but they form the principal stumbling blocks.

First comes the question of the six days of Genesis 1 reiterated (in what must often give pause to modern congregations who are called upon to recite it) in the Sabbath commandment of the Decalogue (Exodus 20:8–11). We have seen reasons for maintaining that the problem here is that the Creator God of the Bible is infinitely greater than the domesticated, easily managed and readily understood deity most objectors have at the back of their minds. For the Bible's Creator more than fills the most exalted conceptual framework we can ever devise for him. In making himself known to men he has had therefore to stoop to enter the rooms of their understanding, and this is the real cause of the problem. As John Calvin, the prince of commentators, wrote centuries ago, the Spirit of God speaking[1] in Scripture 'chose rather after a sort to stammer than to shut up the way of learning from the vulgar and unlearned sort'.[2] That the phased, progressive establishment of the physical universe for the purposes the Creator had in mind is spoken of in terms of a period of *six days followed by a seventh* is not difficult to understand (on the premise of divine inspiration) in view of the lessons the Scriptures proceed to draw from it: the repetitive weekly pattern of life designed for human well-being;[3] the divine imposition[4] of a cyclic pattern on so much that belongs to natural process;[5] and the brevity of all material existence compared with the eternity of God.[6] All these emphases would have been missed or greatly weakened if (inconceivably) the Bible's record of origins had been given in the sort of age-long, evolutionary terms in which objectors seem to think it should have been given.

Next we recall that the tendentious concept of 'special creation' (in the sense we have earlier given it) cannot be laid unequivocally at the door of Scripture. The biblical language is, in fact, extremely open in its description of how life-forms actually appeared on earth. The earth 'vegetated vegetation'; the waters 'swarmed with swarms of living creatures'; the earth 'brought forth living creatures'.[7] Further, the most obvious meaning of the statements in Genesis 1:11,12 containing the key phrase 'according to its kind' is that each kind of plant bears it own kind of seed, and this is indeed no trivial statement. It asserts a very fundamental matter—the *contingency* on the will of the Creator of the order we know so well. Things might conceivably have been otherwise! The phrase has an eye too to the great variety in plant life, an emphasis even more evident in the case of the marine and land animals (Genesis 1:20,24), where the Jerusalem Bible renders the text 'God created . . . *every kind of* living creature with which the waters teem . . . God made *every kind of* wild beast, *every kind of* cattle, and *every kind of* land reptile.'[8] Living creatures in rich variety came from the Creator's hands! Either of these distinctions of meaning removes any cogency from the argument that here we have a definite statement of the 'fixity of species' or 'Special Creation', any more than we have in Psalm 93:1 a definite statement of the mechanical fixity of the earth.

We go on to recall that the biblical data give us strong reasons for believing that the primal creation was not (as is often imagined) idyllically perfect. It contained fierce animals; death and predation were among its features. For man was to subdue and civilize it; his was to be a Messianic role. The coming of man, a creature 'made in the image of God' to have fellowship with God, was accordingly associated with the culture of plants, in a selected environment. One of his tasks, it seems likely, was to bring about a vegetarian basis for animal nutrition as well as for his own; but it is nevertheless quite consistent with the biblical data to believe that he himself began as an eater of meat as well.

The first man of whom the Bible speaks, Adam, was formed (materially) of dust from the ground, as were the land animals and birds.[9] But in the case of man (alone) a further detail is added: God breathed into his nostrils the breath ($n^e\check{s}\bar{a}m\hat{a}$) of life. This seems to indicate for man a unique relationship with the Spirit of God, not shared with the animals. It recalls the incident recorded in John 20:22, when 'he [Jesus] breathed on them and said to them, "Receive the Holy Spirit",' an incident which undoubtedly looks back to Genesis 2:7.[10] This is significant, for the New Testament there speaks of a new creation (among men) differentiated from the old in just this way i.e. in its relationship to the Spirit of God. This point will be taken up again very shortly; but for the moment we note that the Bible does not positively imply that this creative act constituted Adam the physical ancestor of the entire human race. Rather, it established him as the racial head, the 'type' of Jesus Christ. This is an interpretation which has been argued at length in Chapter 9. Outside the Garden, unbreathed upon, the Bible gives us reason to believe were other members of the race from which Adam had been taken (as millennia later Abraham was to be taken from his compatriots and removed to a new land). From among these one was brought to Adam in the Garden, to the accompaniment of a dream, to be his wife (as later Rachel was to be brought to Isaac). The purpose of Adam's segregation, we may surmise, was to equip him for leadership in fulfilling man's role of 'subduing the earth . . . and having dominion'. Certainly at some time he or his descendants would have had to leave the confines of the Garden if man was to populate the whole earth; the residence there was only a temporary expedient. There are biblical parallels for all these suggestions therefore.[11]

Adam failed in his calling, as later the nation of Israel failed in its.[12] What was the result? 'Sin came into the world and death through sin'.[13] Adam and Eve, having chosen moral autonomy instead of obedience to the Creator as their principle of conduct[14] were expelled from the Garden. They lost the privilege of familiarity with God symbolized by the tree of life; they were 'cast forth as a branch and withered' (cf. John 15:5,6). This death of the spirit in alienation from God constitutes the 'sting', that which now makes the prospect of an end to our physical life so sad and oppressive.[15]

Much of this we have expressed in theological terms. The question must now be faced: how can it be made to tie-up plausibly with events which may have happened actually (and in essence, observably) on the plane of human history? This is a legitimate and pressing question, and to answer it we must undertake somewhat of a digression. To begin, suppose an infant a day or two old to be abandoned on a desert island and to be almost miraculously saved from death by an animal mother who reared it with her own litter. Twenty years later such an individual could not rank as fully human.

Its personality would be quite undeveloped; its ability to communicate rudimentary;[16] its self-consciousness quite questionable;[17] its moral sense limited to what was, in an animal way, merely anti-social. It could hardly be blamed if it walked off with a visitor's watch! It would have all the genes; but what the genes were capable of mediating in the way of truly human personality would remain largely unrealized. The case, in fact, illustrates what we mean by human solidarity: men and women cannot be fully human apart from contact with other men and women.

When Charles Darwin first encountered the natives of Tierra del Fuego he was much impressed with their low development:

> The Fuegians are in a more miserable state of barbarism than I had expected ever to have seen a human being. In this inclement country, they are absolutely naked, and their temporary houses are like what children make in summer, with boughs of trees. . . . I shall never forget, when entering Good Success Bay, the yell with which a party received us. They were seated on a rocky point, surrounded by the dark forest of beech; as they threw their arms wildly around thier heads and their long hair streaming they seemed the troubled spirits of another world.[18]

> But, I have seen nothing, which more completely astonished me, than the first sight of a Savage; It was a naked Fuegian his long hair blowing about, his face besmeared with paint. There is in their countenances, an expression, which I believe to those who have not seen it, must be inconceivably wild. Standing on a rock he uttered tones and made gesticulations than which, the cries of domestic animals are far more intelligible.[19]

Many years later (1870) Darwin heard of the work done by Christian missionaries in Tierra del Fuego. He was amazed at the change brought about in the natives. Darwin wrote to a fellow shipmate of the *Beagle* days, now Admiral James Sullivan of the South American Missionary Society:

> . . . the success of the T. del Fuego mission. It is most wonderful, and shames me, as I always prophesied utter failure. It is a grand success. I shall feel proud if your Committee think fit to elect me an honorary member of your Society.[20]

Darwin, as missionary work later showed, had been mistaken about the Fuegians. They were far more 'human' than he had imagined, with a complex language and involved religious beliefs. No matter; they still serve as illustration of an important point. The unevangelized Fuegians, like the infant reared by an animal mother, had potentialities which for different reasons had (conceivably) never been actualized. So far as we know, their condition had never been higher than it was when Darwin saw them. What transformed them was contact with men and women who taught them ennobling things. No doubt outside contact taught them also undesirable things—how to be worse sinners as well as greater saints. That goes without saying; but it reinforces the point being made: the profound effect of what the secular historian would call 'culture'. The matter can be applied in this way.

Imagine a stock of primates living together in a primitive society, and never having known anything higher. Among them come a missionary and his wife. What will happen? There might, on the one hand, be a dramatic rise in the level of their culture, materially and spiritually. Darwin learnt and acknowledged this in the case of the Fuegians. On the other hand, nothing might happen. The primitive society might prove incapable of elevation. The great apes exemplify this. Why the difference? It might be said to lie in a latent potentiality, present in the Fuegians (but never before actualized), absent in the apes. Now, it might be urged, doesn't the presence or absence of this potentiality constitute a difference so distinct and discontinuous that we must regard it as a matter not of degree but of kind, and rule out any continuous developmental connection between the two? Not at all; and we can demonstrate this by an elementary argument.

Consider a simple system—very simple indeed, a mere mixture of the two gases nitrogen and hydrogen.[21] Suppose this mixture to grow progressively richer in hydrogen. There comes a point, more or less suddenly, when it acquires a striking new property—the ability of a jet of the mixed gases to sustain a flame. Before this point, expose it to a spark and nothing happens; after it, the thing is alight! This easily comprehended example shows that there should be no difficulty whatever in supposing that the progressive enrichment of the material system mediating life can result in the acquisition, more or less suddenly, of a completely new possibility never dreamed of before. For 'flame' read 'spirit', and the case we are arguing, if not quite proved, is rendered very plausible.

We can now return to our previous train of thought. How can the biblical story of Adam and Eve be woven-in to the fabric of the presumed evolutionary origins of our race? Possibly in this way. A hominid stock had evolved to the point where the potentiality for true humanity had appeared. The members were doubtless more primitive than the Fuegians Darwin saw, but they may have had a rudimentary language, some power of conceptualisation, musical and artistic sense, tool-making ability, the power to plan ahead, a recognition of and an elementary response to death.[22] But the members were still not truly human; the flame of true humanity had not yet been lit. One of their number—Adam—was taken aside; into his nostrils was breathed the *nešāmâ* of life; he was instructed in agriculture;[23] and he was sent back to his people—not, alas, (as it turned out) to teach them exclusively the good life of authentic humanity, but to initiate them as well into 'the knowledge of good and evil'. The penal nature of the 'death' which Adam had come to know because of disobedience he could not but communicate to his race. That, and how, God had expressed his will for man was now a matter of common knowledge, and experience became for each of them what later it was for the aspostle Paul: 'Once I was alive apart from law; but when the commandment came, sin sprang to life and I died'.[24] The mere overstepping of the social code of an animal existence had now found a wholly new and burdensome theological dimension: it was *sin*, an offence to God. The innocence and bliss of ignorance had gone forever.[25] Simultaneously, physical decease had acquired a deadly sting; it was the painful anticipation of judgement to come.[26] The solidarity of the race (as defined earlier) had seen to it that the penalty extended to all.

Doesn't a suggestion along these lines make sense first, of the biblical data; second, of man's inner experience; and third, of the scientific evidence? I believe it does. And it makes immediately comprehensible Paul's assertions that 'by one man's disobedience many were made sinners'; that 'one man's trespass led to condemnation for all men'; and that 'death spread to all men because all men sinned', even if their sin was not, like Adam's, the disobeying of a direct command.[27]

No doubt at this point (if not before) an objection will be raised. 'This all sounds very nice', it will be said, 'but isn't it going far beyond what the Scriptures tell us—particularly what the Old Testament does? Isn't there too much speculation here to be acceptable?' I admit the force of this objection, but I think it can be overestimated. I would be much more cautious of speculation were it intended as the foundation for any considerable dogmatic edifice. We can be sure the Holy Spirit would have given us something much firmer for that. But the purpose here is something far less ambitious. It is simply to remove difficulties—to show that the data of revelation and of science are not irreconcilable. There is surely a legitimate place for speculation in this. Further, the speculation has been guided and controlled by explicitly biblical principles—the well-known Pauline analogy of Adam and Christ, and the parallel between Adam and Abraham, each the father of a 'multitude of nations'. Finally, there is the general recognition that it is part of the genius of the Bible to tell its story very selectively, even in what might be called a fragmentary fashion. Of none of its great characters (even of our Lord) are we given a biography anywhere near adequate by modern literary standards. Its history is confined to Israel, and even then is extremely sketchy. It passes over in almost complete silence the outstanding problem of the origin of evil; and so on. The Preacher refuses to be side-tracked; his one purpose is to teach men and women how to live in accordance with the will of God, and he never deviates from it.[28]

We conclude this chapter with a brief recapitulation of some points of rather more technical concern. The inherited variations which are the raw material of natural selection are almost universally held to be random; unrelated that is to the organism's needs or to the direction evolution is taking. This somewhat limited use of the idea of randomness asks for less than would be implied by full randomness in the physicist's sense. But even if it were true that variations were random in the latter sense *it would still not follow that they were outside the directive control of the will of God,*[29] that is, of the Bible's God. It is a fallacy to reason therefore that chance, established as such by any statistical test open to us, negates God. It negates only the small gods of secular thought.

But that inherited variations are biologically random is a proposition which is by no means certain. It has been questioned often in the course of the purely scientific debate on evolution,[30] and it is still being questioned. It is probably quite impossible to substantiate it; and where it is maintained it can be maintained only as a methodological principle. In so far as it proves inadequate to explain major phenomena (such as the origin of great groups like the mammals or angiosperms) it points to an alternative possibility: that inherited variations have not (or not always) been 'random', but that at least at certain great moments in the history of life they have been positively directed into predetermined channels by some agency scientifically unidentifiable, resulting in the rapid appearance of major new forms. This may look to some suspiciously like special creation. Perhaps; but it would account for the phenomena noted. This suggestion is not a mere revival of the vitalist's idea of entelechy or of Bergson's *élan vital,* both of which refer to a principle intrinsic to nature. It invokes rather the special working of the transcendent Creator; special in the sense that the pattern of his activity at these points lies outside of that which he normally follows. But it must be admitted that it would be vain to try to establish this suggestion by scientific criteria as some 'creationists' seem to be doing. God's ways in nature are essentially untraceable, the Bible tells us.[31] All we would find if we attempted to push our enquiry to the bitter end would be that our methodology, like Balaam's ass, lay down and refused to take us further. Already there are suggestions that physics has reached this point.[32] If the universe is truly contingent, then we are bound ultimately to reach it, and the suggestion we have been discussing is merely that the biologists will reach it sooner rather than later. God's inscrutable activity will thrust itself upon them earlier than they expected. Let me appeal to my scientific colleagues: is it really inconceivable that we shall run up against such a methodological dead end? Surely not. Then why not open the mind to the majestic biblical declaration: 'In the beginning, God created . . .'? There is every reason to do so.

CHAPTER FOURTEEN

Creation—How shall we think of it?

The doctrine of creation as it appears in the Bible has been very inadequately studied and as a consequence is widely misunderstood. The naïvety of the language of Genesis misleads minds accustomed to buzz words and studied profundity. True profundity can afford to be simple; indeed it prefers to be.

Faced with a problem in understanding, the customary reaction of the mind is to try to find a model in terms of which to think. Accordingly, a model is proposed to illuminate the Bible's doctrine of creation. With its help the main features seem to fall into place. In particular, it appears that the doctrine of evolution may find a subordinate place within the Bible's overall teaching.

Darwinism in particular and evolution in general are subjects which continue to provoke a continuous stream of literature, some scholarly and some not. A recent example by an able writer on the secularist side[1] illustrates an all-too-common failing of such literature: the hasty assumption that the biblical doctrine of creation has been adequately grasped and so can be confidently criticized. Thus R. W. Clark refers to the Victorian belief that 'living things . . . were, *as Genesis maintained,* a pyramid of immutable species' with man at the top,[2] the phrase I have italicized evidently (from his later remarks) representing his own view of Genesis and not merely that of the Victorians. The tendency to assume that the Genesis account is so simple and artless that anyone can comprehend it at first glance appears to be very widespread; probably even Darwin himself erred in this direction. It certainly seems to be the case that compared with the vast amount of time, concentration and devoted energy that he gave to the problem of the origin of the species the effort he gave to understanding the Bible was quite negligible. His biological work doubtless largely monopolized his attention. In later life he wrote: '. . . for many years I cannot endure to read a line of poetry; I have tried lately to read Shakespeare and found it so intolerably dull that it nauseated me. I have also almost lost any taste for pictures or music.'[3] These are alarming, indeed ominous words from such a thinker, a man whom many moderns regard as having unearthed the

115

real clue to man's nature! But what had happened to this very gentle and lovable man? The answer seems to lie in a remark of his own: 'The habit of looking for one kind of meaning I suppose deadens the perception of another'.[4] For unless we happen to have inside information that (beside the scientific one) there *are* no other meanings worth bothering about, this remark of Darwin's must be recognized as indicating an insidious and injurious attitude of mind, ever threatening the dedicated investigator. Great man as he was, Darwin fell a victim to it.

It is one of the theses of this essay that the biblical doctrine has been widely misinterpreted and misunderstood, by both believers and unbelievers. Particularly has this been so on the level at which it seems *prima facie* to have implications for science. The understanding of the 'six days' and of the phrase 'after their kind' are outstanding instances of this. Why has it been so? There are doubtless more reasons than one,[5] but there is one reason worth pausing to consider. It is particularly influential at the present time, and it stems from the fact that the style of Genesis seems very naïve. Thus, God works on six days and rests on the seventh. He forms man of dust from the ground, and plants a garden of lovely and appetizing trees for man's sustenance. Adam gives names to the animals; Eve is tempted by a snake and takes the forbidden fruit. Adam and Eve are ashamed when they discover their nakedness and try to hide it with an apron of fig leaves, and so on. Even to a child the narrative evokes vivid and immediate pictures. Now we live in an age when there is an explosion in communication. More writers than ever have something they want to say, and to say publicly. Means have multiplied for doing so, from books and periodicals to photocopiers and letters to the press. Unfortunately not every writer has something worth saying, or can say his piece well, or is free from motives of self-advertisment in saying it. These things, as much as any, have resulted in the spread of the style sometimes referred to as 'gobbledegook', rich in buzz words and phrases, circumlocutions, ill-conceived assertions, and bogus profundity. (There is a similar phenomenon in art and music). Concurrently, there is a multiplication of books for children. But children don't like gobbledegook. They can't understand it and are not susceptible to the appeal of the 'emperor's clothes'. So children's books are written in plain and simple language. Now put these two circumstances together: eager and thoughtful adults all too often finding themselves faced with scientific, philosophical or religious writing too deep (apparently) to understand; and children revelling in what is expressed with charming simplicity. Is it any wonder that the idea has got about that what looks naïve must be child's stuff, or, if written for adults, suitable only for primitive society? That the last thing to look for in it is a serious challenge to thought, or a call to re-evaluate one's existence? Surely not. Yet it is an idea damagingly prevalent—and often quite false.

This is a sad state of affairs. The truth is, that when an author has something profound to say, the more simply he can say it the better; and the more masterly his grasp of his subject, the more competent will he be to do so. Add to these qualifications a complete dis-interestedness—a lack of desire to impress or to build a reputation—and the result is almost a foregone conclusion. This is the clue to the 'naïvety' of Genesis, as it is to the 'naïvety' of those incomparable words of our Lord, 'Look at the birds of the air; they neither sow nor reap nor gather into barns, and yet your heavenly Father feeds them . . . Consider the lilies of the field . . .'6 Both discourses are far-reaching in their theology, and all-embracing in their significance for life. Their very naïvety is an index of their greatness and gravity.7

We return now to the question of understanding the Creation story. Faced with any novel subject matter outside familiar experience our minds instinctively ferret about seeking for a model in terms of which to think. This isn't necessarily done consciously; sometimes a model springs almost at once to our notice and we hardly recognize it as such. At other times a great deal of laboured thought may go into the process. A well-known example of model-making in Science (where models are legion) concerns the atom and its structure. Atoms can (it was discovered) emit electrons, minute particles of negative electricity. This was something out of the ordinary; how should we visualize it? Electrons must, of course, be balanced by a positive charge. How are the two related to form the neutral atom? First came Sir J. J. Thomson's plum-pudding model, with the electrons as sultanas; then Lord Rutherford's solar-system model, with the electrons as planets; then the Bohr and later models, with the whole thing now highly sophisticated. This was all model-making realized as such, but it illustrates how our minds habitually work, whether deliberately or not. There is a sort of inevitable logic about it; the unfamiliar must be understood (and described) in terms of the familiar. How else could we live with it?

The Bible's subject matter is no exception to this rule. Like that other book of God, Nature, it presents us with data on matters outside everyday experience, to understand which we willy-nilly propose models to ourselves. When it tells us, 'God said, "Let there be light"; and there was light', the imagination sets to work at once. Many readers will find themselves thinking in terms of the near-instantaneous flood of illumination when a switch is thrown. When it says, 'The Lord God formed man of dust from the ground, and breathed into his nostrils the breath of life; and man became a living being', they will quite likely be entertaining the thought of a clay modeller who lovingly caresses with his warm breath the little figure he has shaped; and so on. Yet neither of these 'models' is positively demanded by the biblical data,[8] any more than the plum-pudding model was positively demanded by the physical. Yet the latter model was widely held at one time. It was dislodged only when further study proved it to be inadequate. It is, in fact, never surprising if the advance of knowledge necessitates the revision of models; the surprising thing would be if it didn't.

So we come to the main point of this chapter. Is it possible, while remaining wholly loyal to Scripture, to devise a model which will do justice to our growing knowledge of it? Let us try. The subject matter is 'Creation', nothing less than the divine action by which the totality of things was brought into being and continues in being. In the Bible the verb 'create' is used only of God and never of man. However, the Bible does tell us, in the very context of creation, that man was made 'in the image of God'. This would seem to encourage us to look cautiously, in our search for something suitable, to what we commonly call the 'creative arts'. Very well; but among the creative arts what shall we choose? The art of the painter or sculptor or architect pass in review, but they seem too static. The art of the composer has obvious movement, but it lacks other dimensions. Perhaps the best would be one we have already used—the work of the serious dramatic novelist. This comes nearest, probably, to touching human life at all points. It is something too which involves creation, quite literally, through the word. What is created is spoken into existence. Let us follow out the analogy.[9]

Firstly, look broadly at the proposal. The serious dramatic writer and the literary masterpiece on which he is still engaged: are there any damaging objections to entertaining this as representing (in terms we can grasp) the Creator and his on-going world? Of course, any model must fall short; if it didn't it would be the real thing.[10] However this model invites no serious objections, nor does any better one readily suggest itself. So we will proceed.

The model illustrates at once the Bible's insistence—emphatic but not always easy to grasp—that the Creator is both transcendent and immanent. This two-sided doctrine constitutes the difference between theism and the rival ideas of deism (on the one hand) and pantheism (on the other). Transcendence? The author is there, prior to the narrative. He makes up his own mind when to write, what to write, and when to lay down his pen. If he wishes, he can begin another story, of a different kind. Immanence? At every moment in the narrative, conceived as *present,* the author is there, writing, and the story lives and moves only as he works. He holds its world in being, as an on-going thing; its flux is the substance of his animated thought. Very significantly its future is 'open', unpredictable, not wholly determined by its past, a feature highly characteristic of the present scientific world view.

Secondly, this model allows the Creator to inhabit a sphere dimensionally richer than the one he has created, and now holds in being. Clearly this is true of our author. His narrative is only a thread, as it were, drawn out of the fabric of his own larger world. Were this not so it would be impossible for him to write other and different stories! At this point we may remark that the space and time in which the narrative's action moves bear no necessary relation at all to those in which the author is actively at work; science fiction should have made that plain enough. It does not compel us therefore to interpret the six days of creation as periods of *physical* time either long or short. To change to a musical analogy: Mozart wrote his last three symphonies in six weeks. But this period bears no relationship whatsoever to the duration of their actual performance in the concert hall. As a problem therefore the 'six days' ceases to be insuperable, or even threatening, once the model is accepted.

Thirdly, the author model makes sense of *creatio ex nihilo*. For a human author, to be sure, such *creatio* is only doubtfully possible. Man can (arguably) create only from materials drawn from his own given experience, but even so his creative powers—witness mathematics and music—are wide enough. But we need suppose God subject to no such limitation. Even a human author can write of things no one has ever experienced or thought of before (Time Machines?); and some popular writing is extraordinarily imaginative. No; in the sense in which the term is used in biblical theology, *creatio ex nihilo* is certainly not a nonsensical idea.

Fourthly, the model we are discussing seems to do some degree of justice to the biblical notion of time. The world is neither static, nor is it something cyclically never-ending. It manifests a *historical* process; it is going somewhere; it is moving to a consummation. It is, in fact, a story! Time has an arrow on it. Further, without reverting to the cyclical idea, it provides that the world can still be a *created* thing even if it is infinite in time (in both directions). There is no reason why the eternal Author should not have been always writing, or should ever intend to stop. On biblical grounds one might say that he has started a story which he will bring to a climax, and then begin a sequel in a new setting; but the point I wish to make is that the analogy shows that the biblical idea of creation does not stand or fall with the triumph of the Big Bang Theory of Gamov (1946) over the Steady-State Theory of Bondi, Gold and Hoyle (1948). God remains

Creator whichever theory is finally validated. This is a conclusion which (anachronisms apart) was long ago emphasized by Thomas Aquinas (c. 1225–1274). One other important consideration relevant to time remains: the Bible's teaching that God both finished his creative work and yet continues it. Clearly, we can accommodate this by supposing that our author first sets the scene and introduces the *dramatis personæ*; after that, the story (we might almost say 'history') may proceed.

The final point of correspondence between the author model and the biblical doctrine is an important one. It concerns the impropriety of speaking of a 'mechanism' of creation.[11] We saw that the Bible never does so. Nor does it associate the act of creating with any material process in space-time. In so far as it is right to speak of a mechanism or process associated with creating, everything must be regarded as taking place within the divine Mind. Then comes the Word, and the Creation is there! What better illustration of this could we possibly have than authorship?

So we come to apply our model to ourselves: how did mankind arrive on this earthly scene? Was it by evolution or by creation? For concreteness, we think of our favourite novel: *Gone with the Wind, Wuthering Heights,* or *The Forsyte Saga.* We open the book at page 291 (say). We meet there the principal character, alive and kicking. How did he arrive on the scene? For answer, we can turn back the pages one by one and find the occasion on which he first entered the story; then trace the vicissitudes by which he has developed and matured to what he is now. So there we have it! He has *evolved,* and the story is the account of his evolution. But, of course, there is another answer (and one really more fundamental). He was thought-out by the author, created in his mind, and then introduced into the story. Our hero is the *author's creation,* wherever we meet him, from his first appearance to page 291 and beyond! And that is clearly the *ultimate* answer to the question: 'Where did he come from?' 'How did he get there?' Were it not for that, and for the act of writing, there would be no story, and no hero to enquire about.

Does this help? I ask the question of the man or woman who cannot ignore the biblical testimony, and yet who in honesty feels that Darwin has a case. I think it should; in fact, I believe that the author model removes, perhaps completely, the apparent antipathy between biblical revelation and scientific discovery. The question remains, 'Is the model valid?' The secularist may not think so. He gets on very well without the 'hypothesis of God', so why introduce it? Up to a point we can agree with him; the story is interesting and well-written and we can enjoy it without even enquiring whether it has an author. But at this point our analogy leaves us in the lurch. We are not only readers of the story, studying it from outside as it were (this is the purely scientific stance); we are actors in it, members of the *dramatis personæ*. And we cannot opt out. Both Christians and secularists agree so far; but the Christian goes farther. The Author of the story, the Bible tells us, 'made us for himself'. He offers us, through Jesus Christ our Lord, a place in a new and happier story. It is called, 'The Kingdom of God', because God himself takes part in it, as King. He is no longer merely Author! His servants serve him, and they see his face, and the story is everlasting joy![12] This is what secularism rejects; this rejection is its tragedy and its hopelessness. Is there any compelling reason for such rejection? I must leave this question to the personal decision of my reader.

Epilogue

In his Gifford Lectures, published in 1974 as *Animal Nature and Human Nature,* W. H. Thorpe, an animal ethologist of international repute and himself a religious man, wrote as follows:

> '. . . the biblical myth of Adam and Eve and their fall from grace . . . cannot, except by outrageous chicanery (sophistry), be made to accommodate these new insights . . .'[1]

This is not, I believe, a contemptuous dismissal of the biblical testimony; but it is nevertheless a profoundly mistaken one. In what has gone before I have tried to set out my reasons for maintaining this. Prof. Thorpe ends these lectures by indicating where he pins his hopes: on 'natural theology', the 'only possible basis' for which is 'what we call scientific'. Many other eminent thinkers to-day have adopted a similar stance. But what does this stance fundamentally imply? It implies that an understanding of what is of 'ultimate concern'[2] to man must come through his own efforts. It must be painstakingly built up by the use of those faculties with which man finds himself naturally endowed. In stressing this it puts a premium on intellectual gifts. Those without them must tag along behind those with them and gain whatever knowledge they can of what is of 'ultimate concern' from intellectuals who may be themselves even quite irreligious (as Darwin came to be, and as more recently Jacques Monod and Francis Crick have professed themselves). Against this is the stance of historic Christianity: God can be known only by revelation.[3]

Clearly, underlying these two stances are two profoundly differing conceptions of God. The first must picture him as almost casually indifferent to the heartache of lesser men and women who long to know him, but who cannot, alas, achieve the lofty scientific insights through which such knowledge must come! Centuries must elapse before some of them benefit from biologists having discovered natural selection or physicists the anthropic principle; perhaps even after such discoveries these concepts will be beyond their intellectual grasp! Historic Christianity, however, gives a fundamentally different picture of God. He is deeply concerned with the common man. He

122

has, of deliberate choice, 'hidden' the things of ultimate concern 'from the wise and understanding and revealed them to babes'.[4] This is our Lord's testimony. Paul's witness is the same. In his wisdom, he says, God has decreed that knowledge of himself shall come not through human science and philosophy but through the 'folly' of what is revealed.[5] Let me ask the reader: which conception of God answers best to the word 'Father' which our Lord bid us use of him? Surely that of historic Christianity, which makes revelation, not science and philosophy, the key to understanding ourselves and our *'summum bonum'*. If God has not disclosed the ultimate answer to ordinary mortals as such, but left it to be discovered by the accumulated labours of generation after generation of gifted intellectuals, then he can hardly care much for the common man. I believe the conclusion is as simple as that—unless indeed these matters of 'ultimate concern' are really of minor importance, or unless there be no God to reveal them.

Let us imagine that the reader agrees so far, and is ready to accept that the Creator might be reasonably expected to have communicated by revelation his purposes to man. Does the Bible stand up to modern knowledge as such a revelation? Certainly it does. But, a reader may object, what about the dinosaurs? There is surely no difficulty in fitting them into the primal creation. What about the cruelty of natural selection? This has probably been greatly overstated, but it forms part of the general problem of theodicy, and I refer my reader to what is said in Appendix 9. Are there perhaps further difficulties? There may well me a number which I cannot anticipate. But before leaving the matter I would like to ask the reader a question of my own.

All of us live by some sort of philosophy of life. It may be long-pondered or thoughtless, simple or complex, avant-garde or ancient in its wisdom. But it is there; and it has a starting-point, somewhere from which it is prepared to venture forth in its quest for understanding. Perhaps this 'basic presupposition' is the ultimacy of matter and energy? Or that man's mind is the measure of all things? Or that existence is finally meaningless? Or that all is illusion, a bad dream? May I invite the reader to identify and name his or her own starting point?

The Bible presents us with one, and a very convincing case can be made out for this being none other than Jesus Christ. It is from him that our thinking is to set out. 'In the beginning, God . . .' is how the Old Testament introduces things. 'In thy light shall we see light', continues the Psalmist.[6] And these strands are taken up in the New Testament with 'In the beginning was the Word . . .'. 'All things were made by him (came to be through him) . . .'. 'The true light that lightens every man was coming into the world.' 'We beheld his glory,' says John, 'the glory as of the only begotten of the Father, full of grace and truth.'[7] That is where the Bible then bids us begin. If my argument has been sound, that will lead us to the Bible's doctrine of Creation, and to much else. Not to mere matter, lifeless and unfeeling; nor to mere mind, an apparent latecomer anyway; but to Jesus Christ, 'in whom all things hold together'.[8] This is the starting point for the Bible: can the reader suggest a better?

Appendixes

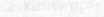

Approach to Genesis

(i) *Can Scripture be used to prove Scripture?*

It is sometimes urged that the appeal to Scripture to prove Scripture (even the appeal to our Lord's pronouncements to do so) is invalid, since it begs the question; it is reasoning in a circle. That this need not be a conclusive objection can be illustrated by reference to a parallel problem:[1] how can we satisfy ourselves of the validity of our physical senses as sources of reliable information? At least *sometimes,* on the face of it, they seem to prove unreliable (witness mirages, ventriloquism and referred pain). If we attempt an answer to this problem it will amost certainly be found that it involves the same sort of apparent question-begging as in the case of Scripture. Thus we may reply that we believe our eyes are not deceiving us becase we can also *touch* the object we see; further, other people can see and touch it too. But this is plainly to assume for our sense of touch (or for other people's senses) the sort of validity we wish to establish for our sense of sight. The same is true of the use of our reasoning faculty. In the last analysis the fact is that by reason of our creaturely status we *cannot* have final analytical proof. Ultimately we have to take a position as 'given', and start from there. The situation is presented firmly and poignantly in Eccles. 3:11 (JB): '. . . though God has permitted man to consider time in its wholeness, man cannot comprehend the work of God from beginning to end.' We shall be wise to accept this, and to ask only what is the most reasonable position from which to start. We must decide for ourselves; but the Bible puts it in unequivocally plain language: 'The fear of the Lord is the beginning of wisdom' (Ps. 111:10; Prov. 1:7; 9:10; cf. Job 28:28; Prov. 15:33).

(ii) *A Testimony*

One of the most scholarly modern writers on the biblical doctrine of Creation is Claus Westermann, Professor of Old Testament at Heidelberg. His monumental *Commentary on Genesis* is in process of being published in English (1985). The following extracts from his *Creation* (E.T. S.P.C.K., 1974) show what a considerable change has come over critical scholarship since Hermann Gunkel wrote in his famous essay of 1895, 'It can be taken as assured that, ultimately, Genesis 1 is of Babylonian origin':[2]

> It is both remarkable and undeniable that the passages dealing with Creation and primeval time which at the high point of the Englightenment had been dismissed as utterly outmoded, have found a hearing once more in the second phase of the technological age. When the astronauts read out the story of Creation from the first chapter of the Bible before setting off for the moon, this was neither emotion nor enthusiasm. Rather, the words of the Creation narrative were suited to the event. In this spirit they were read, in this spirit they were heard by thousands. The achievements of science and technology in the first phase of the technological age gave rise to arguments for questioning the belief in Creation. An achievement in this same area in the second phase provides the occasion for the recitation of the Creation story. . . . The attacks of the Englightenment, with its glorification of the natural sciences and its ridicule of the nursery tales of the biblical–ecclesiastical tradition, have now run their course, and the emotion has evaporated.

> From the time of Kepler, Galileo, and Copernicus right up to the formation of the Marxist materialistic ideology, the Church had no serious encounter with the all-conquering scientific explanation of the origin of the world and of man. One stood by the validity of the biblical Creation accounts and the belief in the Creator. At the same time one acknowledged more or less openly the scientific explanation of the origin of the world and of man. There was, however, no serious concern to build a bridge between the scientific explanations of the world and the biblical account of Creation.

The present essay is a modest attempt to do just that.

Westermann points out the unsuspected universality of many elements of the biblical creation narrative:

> But the limits of the inquiry have not yet been reached. It appears that the Creation traditions of the high cultures of the Mediterranean world have their roots in traditions still more ancient, going right back to primitive cultures. Let one example suffice for the moment: the imagery of the Creation of man out of mud or clay or dust occurs together with the motif of the life-giving breath in Sumerian and Babylonian myths just as it does in many primitive Creation narratives. How is this striking agreement explained?

> The agreement is not limited to striking individual similarities. A

further observation must be made: a consideration of the Creation stories in the context of the account of the origins, that is of Genesis 1–11, shows that the motifs of this narrative are distributed across the whole world. This is most striking in the motifs of the Creation and the Flood which are found in the stories of early times of people on all continents. It is the same with other motifs of the account of the origins, as H. Baumann has shown quite impressively in his book *Creation and Primeval Time of Man in the Mythology of the African Peoples* (1936; 1964). He has set out the African myths of primitive time according to the motifs of the biblical accounts of the origins. There, too, occur the motifs of the first offence, the origin of death, the origin of civilization, fratricide, the building of a tower. With such far-reaching agreements in the motifs, *the earlier explanation of historical dependence is quite inadequate* [my italics]. The starting point must be that the many-sided and distinctive occurrences of the same motifs of the origins, spread over the whole earth, have arisen quite independently. The conclusion is unavoidable that *mankind possessed something common in the stories about primeval time* [my italics]. The narratives express an understanding of the world and of man which in its broad lines and in an earlier epoch was common to races, peoples, and groups throughout the whole world. However much the civilization and thought of the different groups of men have diversified in their later developments, and however broad has become the gap between the different civilizations, there is in the narratives of the primeval periods a common basis of thought and understanding which can have an even further and deeper meaning for the future of mankind. And so the question becomes more pressing: how did these narratives of primeval time arise? Do the narratives themselves allow any conclusions at all as to their origin? Are there still traces of their original meaning?

The answer Westermann gives to these questions does not go so far as I would wish, but it goes a considerable way. They are all, he suggests, a response to man's sense of being under threat in a threatened and threatening world, a sense which was (and remains) of world-wide incidence and power. But why should man, we may ask, as a creature of Nature, feel threatened by Nature, and why should Nature be threatened? These are ultimate questions, and clearly hold a most vital key to the human situation. It is my conviction that there is no satisfactory answer but that given in the Bible: a primeval act of disobedience to the Creator, and the inevitable alienation which this entailed (and entails). It is time we all recognized this, before our civilization destroys itself and perishes miserably.

(iii) *The status of human governments*

The 'higher' or 'historical' criticism of the Bible[3] has been one of the two major influences which have drastically weakened the hold which biblical authority formerly had on the educated mind. When it is accepted that we can trace, in naturalistic terms, just how this body of religious literature has come into being and has been assembled into the canon of 'sacred' scripture, what room is there left for continuing to regard it as in a unique sense God-given, *theopneustos*?[4] Yet the logic behind this reasoning—widely accepted by modern scholars—is quite invalid on biblical premises. It is invalid, that is, from the standpoint of the basic theological teaching of the Bible as a whole. Consider as an example what the Bible says about human government—not human government in general, but human governments in particular. The being-in-power of any particular government can be accounted for by the ordinary principles of demography: a king has *inherited* a kingdom; a strong man has *seized* power; a prime minister has been *popularly elected;* and so on. But in the biblical view, however true such accounts may be on the demographic level (and it accepts their validity), on a deeper level governments exist by the sovereign disposition of God. A particular Pharaoh has been raised up; Nebuchadnezzar and Belshazzar have been set on the throne of Babylon; Pilate has been divinely charged with the administration of Judæa; and so on.[5] What I am arguing is that this attitude to government illuminates the general position of the Bible.[6] However successful critical methods may be in tracing the rise of the biblical literature, if this principle of biblical theology be accepted it remains a valid option to regard the Bible as uniquely the Word of God. Such a claim must of course be judged on the Bible's own merits; with that we can all agree. But countless millions down the ages who have sought to live by its teachings have found those merits adequate.

How God acts in history is one of the great problems of theology; but not to believe that he does so undermines the whole of Christian faith, not just our submission to Scripture as God's Word. As a basic article of belief this principle is already implicit in such a familiar verse as Psalm 127:1, well-loved and universally accepted by Jew and Christian alike:

'Except the Lord build the house, they labour in vain that build it.'

We can believe that a house is built by God even though we have seen tower cranes and concrete mixers on the site. But such belief is *logically* on the same level as belief that a Book is 'given by inspiration of God', even though the experts have traced the whole process of its making. If belief in the former case is possible for the twentieth century intelligence (and it surely is), there is no reason why thinking men and women should not also believe that the Bible remains 'God's Word written'.[7]

APPENDIX TWO

The Creation Week

(i) *The six 'days'*

It has been argued earlier (Chapter 2) that the 'days' of Genesis 1 were not periods of human time (however long or short) but 'days' of the 'life of God'. In terms of the model of Chapter 14, they belong to the 'temporal' dimension in which the author lives and moves and not to the temporal dimension in which the characters and events of the narrative do. This view springs naturally from the Bible itself. However, the literalistic interpretation of the 'days' is so persistent that it is worth enquiring further into its reasonableness. Let us focus our attention on the third 'day'.

The literalist interpretation maintains that the third 'day' was of twenty-four hours,[1] or very nearly so. This means that the 'gathering together of the waters' into seas must have involved phenomenal rates of flow, with miraculously decreased viscosity of water.[2] It must have involved also vastly increased rates of plant growth as the earth 'put forth vegetation . . . and trees'. Now what does it *mean* to say that the third 'day' occupied twenty-four hours? In reply to this question we might adopt an absolutist position. It means that that was how *God* saw it; the spin rate of the earth was the same then as now *in his sight*. However, there is no biblical warrant for this position; the Genesis account, by its own declaration, is *anthropo*centric, not theocentric.[3] So we must seek a different logic. We must find some standard by which the third 'day' might in principle have been measured and found to be twenty-four hours. It would be ill-conceived to appeal to modern instrumentation (like cæsium or quartz clocks) to provide a standard, for this would forget that the Bible was given through human authors and in a form designed to make sense to their fellows. No; the standard must be provided by time-indicating devices familiar in the author's world and in principle capable of having functioned on the 'third day'. The claim that this 'day' was of twenty-four hours can surely be taken seriously only on these terms. But when we do so examine it it collapses at once. Why?

Because of the 'clocks' familiar in the author's world—various kinds of water clock and sundial, the growth and rhythms of plant and animal life, the human psychophysiological sense of time—not one is eligible. Sun clocks, of course, beg the whole question; animals and humans had not yet arrived on the earth; and most significantly, the rate of water flow and plant growth had been miraculously and indefinitely speeded up, playing havoc with any utility they might have had! In these circumstances to say that the 'third day' had a duration of twenty-four hours is to make a meaningless statement, one that could not be conceptually validated. The interpretation in question therefore virtually collapses under its own weight.

(ii) *'Evening came and morning came'* (NEB)

If this phrase was not intended to equate the days with our common day, what function does it fulfil in the narrative? To answer this question we can pose another. What would be lost if the phrase was omitted? Instead of 'So evening came and morning came, a third day' (NEB) we should just have '. . . a third day', or '. . . day third'. The answer to our question would seem to be that we should lose the vivid sense of God's tireless and continuous creative activity; the days become simply time markers, like the numbers on our calendars. The great creation psalm (104) pictures man as 'going forth to his work and to his labour until the evening' (v. 23). The beasts of the forest by contrast, come out by night until the morning, then they lie down in their dens (vv. 20–22). Evening comes, and man lays down his tools; morning comes, and the lions lay themselves down. But God works without a break; evening comes and morning comes, and God the Creator works on for the day is his, and the night also![4] Only when all is finished does he cease (Genesis 2:2,3 NEB, GNB). This seems to be the thrust of this striking phrase, not the identification of the days as periods of twenty-four hours. Understood thus this phrase fulfils an important purpose. It is meant to set before us a characteristic of the Creator's activity we are to emulate—hard work and solid application till the task is finished.[5]

(iii) *'Thus the heavens and the earth were finished.'*

This leads to a consideration of what the Bible means by 'finished', especially in view of its teaching that creating is a continuing activity of God (see Chapter 4). On the material level the meaning could well include something like this: matter has inherent limitations. We cannot go on building bigger and bigger bridges; ultimately the strength/weight ratio of the materials available sets a limit. Animals with exoskeletons (like the lobster) cannot increase in size indefinitely;

the diffusion rate of oxygen clamps down on them. All sorts of things encounter limiting restrictions of this sort, and there is no reason to doubt that the computer functions of the brain, for instance, are among them. What Genesis may be implying is that in man this limit has been reached. Given a constitution in which flesh and blood are integral elements, man is the noblest creature possible. With him creation is brought to its climax and completed.

There is a theological meaning more important still to be set alongside this. With man, God had fashioned a creature 'in his own image', able to render him 'reasonable service'.[6] This is the highest possible life open to a created being, the *summum bonum*. In man this culmination has been finally reached. But the work of creating continues still. There is a biblical parallel to this in the reconciling work of Christ. This was finished on the Cross;[7] but the work of reconciling men and women goes on.[8]

(iv) *Hebrews 11:3*

'By faith we understand that the world was created by the word of God, so that what is seen was not made out of things which appear' (RSV, RV).

This important passage is apt to lose some of its force in the familiar versions. The 'world' ('universe' NIV, NEB) is not the usual *kosmos,* but *aiōnes,*[9] the world under its temporal aspects. Again, the word rendered 'was created' is not the usual one but is from *katartizō* which means basically 'to render fit', complete; to prepare'.[10] It is *by faith* the verse asserts, that we understand, perceive with the mind (*noeō*), that what we see with our eyes (*to blepomenon*) has not come into being (*gegonenai*) out of what is phenomenal (*ek phainomenōn*); the existence of what is physically seen is not explicable ultimately in terms of physical antecedents. Its source lies elsewhere and deeper, in God's *fiat*. From this it follows that creation is not a truth which can be established (or for that matter, challenged) by the data and methods of science, for these are limited to the phenomenal. It can be known only by faith, i.e. through revelation received as such. This is a most important biblical emphasis, often repeated in Scripture in other contexts.[11]

We may go on to ask why the writer refers to the world under its *temporal* aspects; what are the 'aeons' which have been 'completed' and 'prepared'? They might be successive eras of human history, or of God's covenantal dealings with man; but it seems much more likely that they refer to the Genesis Creation narrative (as indeed almost all translators seem to have assumed). This would agree with the immediate transition in the next verse to the story of Cain and

Abel.[12] Thus it seems a reasonable conclusion that the reference here is to the six 'days'. Time is regarded concretely in Hebrew thought and not distinguished sharply from the events which occupy it. Hence the 'completing of the aeons' could be taken naturally enough to mean the successful accomplishing the planned objective of each 'day', the whole constituting the phased preparation of the earth for man. This interpretation seems to make excellent sense.[13]

If this be granted, the fact that the writer refers to the 'days' as *aiōnes* and avoids the common word for days (*hēmera*) is surely significant. It adds its support to the view expressed in Chapter 2, that the 'days' were days of eternity, not of physical time (let alone of twenty-four hours). Paul's words in 1 Timothy 1:17 (literally, 'King of the aeons . . . unto the aeons of the aeons') illustrates such a usage of *aiōnes*.

(v) *Exodus 20:8–11*

This passage, which enjoined the sabbath upon Israel, is sometimes quoted as settling (in the affirmative) the question, 'Were the six days of creation common days or not?' The argument, however, is quite lacking in cogency. Not only does God's own seventh day (with no 'morning and evening') fail to qualify as a common day, but the passage makes equally good (or better) sense if the 'days' are days of God's eternity instead of man's physical time. The argument, in fact, misconceives the significance of the passage. God's work is here set out as the archetype of man's work.[14] Work is a means to an end, it implies, not an end in itself. In the Genesis narrative the work of creation moves to a climax, which for God is the rest of rejoicing in his work.[15] This is to be the pattern of man's life too; its great end is not work, but the sabbath rest of joy in God.[16]

It may be suggested that the Hebrew word *'āśâ* (make) is used of God in verse 11 rather than the more explicit *bārā'* (create) to emphasize further the parallel between man and God. Man is to 'do' (*'āśâ* again, verses 9,10) his own work on the pattern of God's.[17]

(vi) *'In the Beginning'* by Henri Blocher

In this fine, scholarly and thorough exegesis of the early chapters of Genesis Prof. Blocher discusses very carefully the interpretations that have been given by conservative scholars of the 'six days plus one'. The scheme he favours is a satisfying one; he calls it the 'Literary Interpretation'. It has a long and honourable pedigree. Broadly, it regards the 'six days plus one' of the narrative as an artistic arrangement, never intended to be taken literally or as a chronological or 'scientific' statement. Rather, it represents 'the great logical articulations of the divine work . . . the creation is indeed the archetype of human work.' As such, the narrative in its 'six plus one' pattern relativises human work; the goal of creation for man, it indicates, is not *work,* but the *sabbath of communion* with the Creator. This, in fact, says Prof. Blocher, sums up the difference between the biblical and the Marxist visions. Logical and anthropological rather than chronological and cosmological: that is the clue to understanding what the author of Genesis took as his guiding principles when he fashioned his account around the framework of days.

How does this view compare with the one advanced in the present study? At first sight they are quite different, but closer inspection suggests otherwise. In effect, the present one is that the days represent stages in the conceiving of creation in the Divine mind,[18] as an author works out the plot of his story. But since 'the Lord by wisdom . . . by understanding . . . and by knowledge' established the cosmos (Proverbs 3:19,20), it is not a far cry from this view to believe that the work attributed to the successive days follows a logical order rather than a chronological one. If Blocher regards the guidelines set for himself by the author as logical and anthropological, those suggested in the present study may well be called logical and theological. There is not a great difference here. Both interpretations may be valid.

APPENDIX THREE

Creation and Providence

(i) *The relationship between Creation and Providence*

Unlike 'Creation', 'Providence' is not a word found in our familiar versions of the Bible. Nevertheless, as a theological idea it has abundant biblical justification (see for instance Genesis 45:7; Psalms 65:9; 104:27,28; Matthew 10:29,30; Luke 2:1–7 with Micah 5:2; Acts 10:1–23). Indeed, the theme of God's providential ordering of events is extremely prominent in Scripture.[1] This being so, what is the relationship between Creation and Providence? Various views have been put forward: Creation initiates; Providence sustains; Creation is miraculous, Providence follows natural law; and so on. The view suggested here is different. The Bible, it implies, regards any event as originating both creation-wise and providence-wise. Creation-wise, it originates in the mind of God; providence-wise, it is related to its antecedents in space-time. This distinction was illustrated in Chapter 4 by the example of the individual's creation and procreation. The passage 2 Peter 3:5–7 supports it in that it sees the *stable existence* of the world as betokening God's *creative* activity (as its *observable functioning* represents his *providence*). Thus, 'by the word of God', Peter says, 'heavens existed long ago and an earth formed out of water and by means of water', an obvious reference to the creation record in Genesis 1. This establishes that *sunistēmi* used intransitively (and here translated 'formed' [RSV], 'created' [NEB] may refer to an act of creation. Now its only other comparable use in the New Testament is in Colossians 1:17 (where it is translated 'hold together' [RSV], 'are held together' [NEB]).[2] Here it clearly refers to the *continuing existence* of the cosmos. This is presumptive evidence that such existence is here attributed to the creative activity of the Word (rather than to God's providential care). But Peter goes further. Having spoken of the role of the Word in creation (verse 5) he proceeds to say (verse 7), 'By the *same word* the present heavens and earth are reserved for fire, *being kept* for the day of judgement.'[3] This seems to make the point fairly conclusively.

(ii) *'According to its kind'*

The Hebrew word translated thus (*l^e mînāh*)[4] is composed of three elements: the noun *mîn* (kind, species); the pronominal suffix *āh* (its); and the 'inseparable' preposition *l^e*. A critical question in deciding whether the Bible teaches 'special creation' or not is clearly the meaning of the preposition *l^e*. The basic sense of this is 'to, for, in regard to', but its usage is fairly wide; about sixteen columns (eight pages) of the *Lexicon* of Brown, Driver and Briggs are devoted to it. Their section in which the relevant instances figure is headed: 'Of reference to a norm or standard, *according to, after, by.*' Some typical references they quote are (1) Gen. 1:11, 8:19; Numb. 1:2,20; and (2) Gen. 13:3; Exod. 17:1 (my subdivision). The example in Gen. 1 is, of course, the one we wish to understand. The others (which should be consulted in the RV, which keeps very close to the original) are worth examining. The thrust, semantically, of the phrase *'after* their families' in Gen. 8:19, is to emphasize that the exodus from the Ark constituted not an undifferentiated and confused movement of animal life but an organized and disciplined one. Variety and order were its hallmarks. Similarly, in Numb. 1:2,20 (where the preposition, occurring three times, is translated *by* in RV) it is the diversification and well-orderedness that are prominent. In the second pair of references the semantic thrust is similar. Gen. 13:3 pictures Abraham as taking not an undivided journey to Bethel, but as proceeding by a number of distinct stages (compare '*on* his journeys' RV with RSV, NIV). Similarly, Exod. 17:1 uses the construction to the same effect (again compare RV '*by* their journeys' with RSV, NIV). In a further common use (again following the *Lexicon*) the effect is *distributive*. Thus in Isaiah 33:2 '*by* mornings' means simply 'every morning' (RV, RSV).

We may sum up by saying that it is entirely reasonable to interpret the phrase in Genesis 1 as simply but pointedly gathering up all the ordered and varied categories under discussion and affirming that they came-to-be through the creative command of God (cf. John 1:3). This is far from teaching 'Special Creation'.

We can probably argue further. In Genesis 1 *l^emînāh* and its related forms are used in three settings only: (1) after 'God said'; (2) with the verb *bārā'* (create); (3) with the general verb *'āśâ* (do, make) which can serve as a simple literary alternative to *bārā'* (cf. Gen. 1:26,27; 5:1,2). It is not used with verbs (such as *yāṣar*, form) which imply the use of material and process. If the distinction which has been drawn between the biblical ideas of *creation* and *providence* is valid, then the fact that *l^emînāh* is used only in these settings (all viewing origins from the standpoint of creation rather than of providence) is probably significant. 'Special creation' (like evolution) must rank logically as a process, even if it is instantaneous. It seeks to answer the question, 'Biologically, how?'; and its answer can be described only in terms of space, time, and the material biosphere (as the 'scientific creationists' recognize). Thus, if true, it would properly belong to the *providence* story of origins, and as such would not have the benefit of *l^emînāh*, even if that benefit were forthcoming. For this additional reason therefore 'special creation' cannot be established from Genesis.

APPENDIX FOUR

The Primal Creation

(i) *The 'great sea monsters' of Genesis 1:21*

A pointed reference in the creation story is to the 'great sea monsters' (*tannînim*). These are probably introduced here as an anti-mythical thrust to cut down to size the evil monsters who in Canaanite mythology opposed the god Baal in his struggle with Chaos. They are, like everything else, God's creatures, not his rivals; as such, they are the recipients of God's blessing, Genesis asserts.[1] But it does not suggest they were not fierce! That they were, in fact, regarded by later generations not as docile but as savage creatures is evident from their employment as symbols in such passages as Deut. 32:33; Pss. 74:13; 91:13; Isaiah 27:1 and Jer. 51:34.[2] They are probably to be so regarded here. It is surely unlikely that the writer of the narrative would have expected his readers to have mentally inserted 'tame ones, of course', after the mention of the 'sea monsters' simply because he happened to know that at the end the creation would be pronounced 'very good'. If 'lions and tigers' had stood in the text instead of *tannînim* would the reader have made this adjustment? Hardly. Thus the mention of the *tannînim* in Genesis 1 reinforces the view put forward in Chapter 6 that the primal creation was not all sweetness and light; it needed to be subdued first.

(ii) *Common-sense considerations*

Science can be said to proceed on two levels. There is the lower level on which it can be described as 'organized common sense', and there is the higher level on which it is the province of experts and specialists only, a level of great power and sophistication. Scientific knowledge on the lower level may be considered public property, and it would naturally be taken for granted by anyone writing for an intelligent audience. This is justification for introducing here some arguments of a (low-level) scientific nature.

That the primal animal creation must have involved suffering and death can be argued as follows:

1 It is an inevitable consequence of breeding capacity. Look, for instance, at the programme for the fifth day. The impression is strong that multiplication was rapid; but whether it was at present-day rates or not, without predation, calamity or natural death overpopulation would have been expected relatively quickly. What then?

2 It is an inevitable consequence of relative sizes and weights. A larger animal, with the best will in the world, could hardly avoid crushing smaller ones in the course of simple perambulation. Similarly a large herbivore could hardly avoid eating smaller animals (e.g. insects), themselves engaged in plant-feeding.

3 It is an inevitable consequence of different environments for living things—water, air, land. Inevitably land animals would be expected to get drowned, at least occasionally, fish to be stranded, birds to get frozen, in the primal creation.

To deny that these eventualities had any place before the Fall is almost to deny that there was any continuity in the world itself, before and after. It becomes too different to be recognizably the same. This is hardly a biblical position.

Man's Probation

(i) *The Two Trees*

The purpose of biblical exegesis is not to find a way of reconciling Scripture with science, but to find out the meaning of Scripture from clues found within Scripture itself. The futility of much secular criticism of the Genesis narrative arises from failure to follow this obvious rule.

The 'tree of life' is met with again in Revelation 2:7 and 22:2,14, the latter passage (in conjunction with Ezekiel 47:12) indicating that the fruit and leaves of the tree were for repeated use, and that their virtues depended upon the river flowing from the very presence of God. Other references to a 'tree of life' occur in Proverbs 3:18, 11:30 and 13:12. These passages surely indicate not only that the tree of life may be interpreted symbolically, but that it is meant to be. In the light of the passages from Revelation and Ezekiel, to be deprived of the tree of life may legitimately be interpreted as to be 'alienated from the life of God' (Ephesians 4:18), denied access to him (Ephesians 2:18). The parallel symbols of the path of life and the fountain of life, both associated with joy in God's presence, under a variety of images, point in the same direction (see Psalms 16:11 and 36:7–9).

This understanding of the tree of life strongly suggests that the 'tree of the knowledge of good and evil' is also to be understood as a symbol. But a symbol of what? The interpretations of the phrase 'knowing good and evil'[1] have been very many; the one which seems to have the best claim is that which equates it with the prerogative of moral judgement. As Blocher says, 'The knowledge of good and evil corresponds to the ability to decide. It is the prerogative of the king who judges his subjects, and of the father who brings up his son.'[2] To take the fruit of the tree is to assert one's moral autonomy. 'I don't need to be told what to do. I can quite well decide for myself', is the meaning of the act. The forbidding of the fruit was an instruction to man not to 'seek to be wiser than became him, nor by trusting to his own understanding cast off the yoke of God, and constitute himself an arbiter and judge of good and evil'.[3] 'The serpent holds out less the prospect of an extension of the capacity for knowledge than the independence that enables a man to decide for himself what will help or hinder him. This is something completely new in that as a result man leaves the protection of divine providence . . . Now man . . . will decide for himself.'[4] 'The guiding principle of his life is no longer obedience but his autonomous knowing and willing, and thus he has really *ceased to understand himself as creature'* (my italics).[5] One can say at once two things to this: it pinpoints the source of humanity's sickness; and it represents a movement decisively reversed in the life of our Lord.[6]

The significance of this discussion of the two trees for apologetic is that it lays bare a message for humanity of such consummate importance that it justifies the narrative in surrendering every secondary (i.e. scientific) interest to get it across. It is nothing less than that 'the righteous shall live by faith'[7] (i.e. deliberate dependence on God) and that 'faith works by love'[8] (i.e. deliberate obedience to God). Besides these truths, the significance of any scientific data is almost trivial! One can live, and live to the full, without a knowledge of the workings of the digestive system, the circulation of the blood, the expanding universe, or even the wonders of DNA. One cannot live without faith and love.

Understanding the biblical language about the trees as a symbolic way of speaking of something spiritually profound does not necessarily mean that one rejects the interpretation that regards the trees as sacramental, i.e. as physically-real trees divinely invested with a meaning for man which went beyond themselves. When we say that an officer 'saluted the flag' we may mean simply that he expressed deep love and reverence for his country (perhaps by an action peculiarly his own). But without denying this sense we may also mean that he made a physical gesture towards a particular piece of cloth. In that case (i.e. when both meanings are present) the physical action becomes a sacramental one. However, we interpret them (sacramentally or as pure symbol) the two trees in the centre of the garden (i.e. with a significance central to life) form an interesting parallel to the sacraments of the New Testament, Baptism and Holy Communion. Eating of the forbiddent tree was a once-for-all assertion of independence; submission to Baptism is a once-for-all lowering of the rebel flag. The tree of life was to be the continuously-available source of life and health; the bread and wine are symbols of the same. In so far as the gospel reverses (and more than reverses) the results of the Fall, this parallelism supports the interpretation of the two trees discussed above.

(ii) *The meaning of death*

'In the day you eat of it you shall die' (Genesis 2:17). This raises two questions. What is meant by the phrase, 'in the day'? And what is meant by, 'you shall die'? Let us deal with these in order.

(1) Henri Blocher (with whom I find myself in general in almost perfect agreement) argues that the meaning of the judgement is, 'In

that day you will fall under the power of a death sentence'[9]—a sentence the date of whose actual execution remains unspecified. This permits 'you shall die' to refer to *physical* death in spite of the 900 years delay. Blocher quotes the story of Solomon and Shimei in 1 Kings 2:36–46 in support of this interpretation. I can agree that 'falling under the power of a death sentence' is both an extremely serious disability, and that it is imposed at once, on the very day. Nevertheless, the admittedly clear meaning of Solomon's words does not satisfactorily settle that of Genesis for two reasons. In the first place, the phrase, 'on the day', in 1 Kings 2:37 may well be taken to qualify 'know' rather than 'die'. (When the king repeats his words on Shimei's return to Jerusalem—verse 42—the situation is unambiguous). Second, the natural intention of Solomon's words is in any case to convey to Shimei the impression that the moment his disobedience becomes known the king will be down on him like a thunderbolt with no delay! Omit from verse 37 the little phrase, 'on the day', and a good deal of this sense of energy and determination is lost. With this in mind it may be said that the function of the words 'in the day' in Genesis is twofold. First, it emphasizes that the action (of eating the forbidden fruit) is by its very nature final and irrevocable; God is not merely forbidding what may tend to become a bad habit. Second (and correspondingly), judgement will fall at once, without delay, and the judgement will be death.

(2) But what is meant by 'death'? Alan Hayward has argued that 'the New Testament concept of *spiritual* death is never found in the early books of the Old Testament. The only kind of death the ancient Hebrews spoke of was *physical* death.'[10] This would be a serious objection if it were true. But that is hardly the case. In a noble passage towards the end of Deuteronomy[11] Moses gives his last exhortation to the people of Israel. 'I call heaven and earth to witness against you this day, that I have set before you life and death, blessing and the curse; therefore choose life.' What is *life?* It is 'loving the Lord your God, obeying his voice, and cleaving to him' (v. 20). That will mean 'length of days' in the promised land. Correspondingly, *death* is also spiritually defined; it is 'a trembling heart . . . a languishing soul . . . dread . . . no assurance'. Time is burdensome; it is length of days under conditions of slavery in a foreign land (Deut. 28:65–68). One might almost say that *physical* death is a release, a boon! No; the Pentateuch, no less than the rest of the Bible, regards the gravity of death to lie in the spiritual sphere, not in the physical.[12] Is there indeed any reason why these very words of Moses should not be regarded as an exposition (in the light of experience) of just what is meant by the words of the Lord God in Eden? I can see none. If this is agreed they can serve as a definition of what the latter mean by

'death'—a spiritually wretched existence, devoid of the warm response of love to God, of the vivid sense of sonship, and of God's fatherly care, till physical dissolution brings down the curtain on a sad episode of existence.

With this understanding of, 'in the day you eat . . . you shall die', the problem of the 900 years' delay ceases to exist. But another problem seems to intrude itself. What are we to make of Paul's insistence that death entered the world through Adam's transgression? Isn't it plain that he means *physical* death, as nearly all commentators have argued? My suggestion is that the traditional understanding of Paul is right—in part. In part, because Paul is not thinking of death simply as something which man shares with the lower animals. He is thinking of death as the King of Terrors,[13] as the exit-from-physical-life-to-judgement, as a physical termination with a spiritual sting.[14] When he says that Adam brought this into the world of man's experience we must take his conception in its totality. When we do so we see that death *in the purely animal sense* (which is all that Darwin's theory is concerned with) is not *as such* something which Adam introduced. It may have been included, but we are not entitled to insist from Paul's words that it must have been. In making the distinction (between animal death and death as the King of Terrors) we are following good New Testament precedent; compare our Lord's words in John 8:51 'Truly, truly, I say to you, if any one keeps my word, he will never see death.' Clearly, this does not refer to animal death in whole or in part; it refers to death-as-a-prelude-to-judgement.[15]

To sum up in concrete terms: we can think tentatively (and speculatively) of Adam's hominid original as subject to physical death-without-terrors-and-without-sting. When into its nostrils was breathed the 'breath of life' ($n^e š ā m â$) it became Adam, the first man. To Adam was spoken in Eden a word corresponding to Deuteronomy 30:19,20. Had he chosen *life* his earthly existence would have endured for a millennium,[16] and then with a transformed body[17] he would have been translated to a higher existence (as our Lord and Forerunner has been).[18] Such would have been the bliss of his family and friends that they would have felt no pain at his going, only joy.[19] As it was, Adam chose *death*. He began to experience its consequences the instant he sinned. His earthly existence continued for 900 years but its close was marked not by translation but by the grave, and bodily decomposition,[20] and by the sorrowing of those left behind.

Most of this is (as admitted) speculation, but it is biblically-guided as far as possible and does point a possible way to answering constructively some nagging questions.

APPENDIX SIX

The Ages of the Antediluvians

Genesis records that before the Flood men lived to an extremely great age: Adam to 930 years, Seth to 912, Enosh to 905, Kenan to 910 and the longest of all, Methuselah, to 969. Noah himself was 600 when the Flood came, and he lived 350 years after it. These figures are certainly surprising; but coming in a book of the stature of Genesis they are given with a 'painful deliberateness' (von Rad) which should preclude us from dismissing them lightly. One thing seems certain: the geneaologies of which they form part are not intended to be regarded as unbroken chains,[1] enabling us to calculate how long ago Adam lived. That is not their purpose. They constitute not a line, but a dotted line, giving historical concreteness to the narrative, keeping it well 'earthed', and enabling us to follow the course of God's redemptive purposes as they unfold. Thus the narrator picks out ten generations from Adam to Noah, then another ten from Noah to Terah, whose death in Haran marked the call of Abraham. This follows the same pattern which Matthew palpably uses in giving the genealogy of our Lord, which he divides into three periods of fourteen generations each (perhaps for mnemonic purposes). Matthew is clearly giving us a 'dotted' line, and there is no reason to doubt that Moses is doing the same. That Noah should have lived to see Abraham attain his sixtieth year, and that Shem his son should have been alive when Jacob and Esau were born (as the assumption of a continuous line would mean),[2] and yet that there should be no shadow of a suggestion of this in the narrative, is surely quite incredible! Archbishop Ussher's date of 4004 BC for the creation therefore need not be taken seriously.

This leaves us the problem of the individual longevities. Other nations have similar traditions; in particular the Sumerian King List has records of ten 'great men' who ruled before the Flood and whose reigns were of the order of 30,000 or 40,000 years![3] The existence of these traditions is, somewhat strangely, held to throw doubt on the biblical record; but it could equally well be held to lend it support. In fact, the latter seems the more reasonable view; if other nations had no such traditions it would surely be that much more difficult to take the biblical record seriously. However, it is worth remarking that there is sometimes an element of doubt about the meaning of large numbers in the Old Testament. The census figures in the book of Numbers are a case in point, and there is an interesting discussion about these by Gordon Wenham.[4] Until further light is thrown on the question it seems wisest to assume that the author intended the figures we now have (for longevities) to be taken at their face values. The prophecy of Isaiah 65:17–25 seems to imply that the Messianic age will again see such extended life spans. There may possibly be a hint of the change in man's tenure of earthly life in the '1,000 years' and the 'three score years and ten' of Psalm 90 (verses 4,10).[5]

APPENDIX SEVEN

The Flood

Although the Flood is not the subject of the present essay this is a suitable place to comment briefly about the biblical record. An ill-conceived attempt has been made, and is still being made (especially in America), to implicate the Flood with the geological record, i.e. to imply that the fossiliferous strata which are the professional province of palaeontologists were all laid down in this catastrophe. It is a consequence of this view that the earth is very young—a mere few dozen millennia at most. If this interpretation of the record is sound it puts the Bible in embarrassing conflict with the well-established and well-authenticated scientific view that the earth is very old indeed. However the view in question cannot be regarded as exegetically sound for at least two good reasons. First, it magnifies the geological significance of the Flood far beyond anything the record itself warrants. It reads *into,* rather than *out of,* Scripture. Thus nowhere does the Bible suggest, even remotely, that geologically-vast quantities of rock and detritus were shifted by the waters, nor that violent earth movements occurred on the scale required. Second, and more seriously, the view referred to materially displaces the focus of concern of the biblical marrative. For the purpose of the divine Preacher in this sermon (as always) is to give priority to spiritual realities rather than physical ones,[1] and the interpretation in question acts arguably to reverse this intention. There is every reason to believe that the phrasing of the narrative is designed to emphasize the point that when God eventually acts in judgement he does so with finality and completeness; nothing escapes, except the righteous remnant.[2] It is for this purpose I believe, that the waters are said to have 'prevailed so mightly upon the earth that all the high mountains under the whole heaven were covered; the waters prevailed above the mountains, covering them fifteen cubits deep'.[3] There was no possible escape. 'By which means, the world that then was, being overflowed with water, perished'.[4]

Scientifically, and in entire loyalty to the biblical record, the narrative can be understood as follows. The cradle of civilization in which the action took place was the flood plain of the great rivers of the Euphrates-Tigris system, an area about 400 miles long and 180 miles broad (650 km by 300 km). To east and west the land rises to elevated plateaux, and to the north to the high mountains of the kingdom of Ararat (Urartu) near Lake Van. The flooding was caused by torrential rain occurring simultaneously with huge tidal waves from the Persian Gulf, perhaps caused by submarine earthquakes ('the windows of heaven' and 'the fountains of the great deep'). The waters surged over the river plain, covering all human settlements; even the high points of the plain were submerged ('fifteen cubits deep above the mountains').[5] As the Ark was borne up and carried northward towards the high ground of Ararat, whichever way the occupants looked out there was nothing but water ('under the whole heaven'). Eventually the Ark grounded in the foothills of Ararat, a resting place not located with any great precision. The raven found the sodden land to its liking; the softer dove preferred to wait till things were more hospitable. As a wind continued to drive the waters back and the land became dry, the human occupants emerged and civilization began again around a new centre.

This schema may not be the only possible interpretation of the text on the physical level; but it shews at least that the narrative is scientifically credible. Bearing in mind the purpose of the narrator, the 'universalist' language ('all the high mountains under the whole heaven', 'every living thing . . . man and animals . . . and birds of the air died') is no insuperable obstacle—compare the similar universalism of Gen. 41:57; Luke 2:1; Acts 2:5; Col. 1:23. It is the sort of impressionistic language the reader is expected to take in his stride surely, a natural way of conveying the sense of the severity of the divine judgement. Again, as with the story of Eden, the fact that the biblical narrative has a parallel in for instance the Gilgamesh epic of ancient Babylon is no compelling reason for denying its status as divinely-given; the existence of other stories (theologically much inferior) can be taken as evidence *for* quite as convincingly as evidence *against*.

I believe that to understand things in the way suggested is to do justice to the genius of Scripture; to try to interpret them as the 'Flood geologists' do is not. For these reasons I have no hesitation in agreeing with those[6] who see the Flood as a widespread but not universal inundation, obliterating a particular civilisation but nowhere near covering the globe. Rather, it is to be seen as an act of purgation designed (like the wilderness judgement and the Babylonian exile)[7] to preserve a godly line through which eventually the promised Deliverer should come. It had no need therefore to be global.[8]

APPENDIX EIGHT

Does God throw Dice?

According to modern physics, when two elementary particles interact it is impossible to predict the outcome, except statistically. Just as we cannot foretell which individuals will die to-morrow but only how many will in a large population, so we cannot foretell the behaviour of individual particles but only the aggregate behaviour of many. In both cases the individual is subject to 'chance'; its behaviour cannot be known in advance. When Einstein asked the famous question which heads this section he, of course, accepted all this, so far as the human observer is concerned. But he could not accept that even God did not know the outcome before it happened; that chance had the edge even over him. The view taken in this essay is that God not only knows the outcome of each particular elementary interaction, but wills it too; what we call 'chance' does not mean uncertainty for him.

However, the view is often held (on the authority of science and philosophy rather than Scripture) that God has so constituted nature that at its lowest level (that of fundamental particles) there is an irreducible, built-in element of chance. When two particles interact, it is suggested, God himself cannot foretell in detail what will happen. He throws a die as it were, or spins a roulette wheel, to see how the pieces must be moved on. Only the average behaviour of large assemblies is determined by him; individual events are out of his control.[1] When it is realized that the radio-active discharge of a single low-energy electron could by simple instrumentation be made to trigger an arsenal of thermonuclear bombs, it will be seen that this raises serious questions. John Habgood,[2] who takes this view, argues that it is this built-in element of chance that produces novelty in evolution—as if God were lacking in imagination! On the assumption that the Creator has a goal in mind, his view would seem to imply[3] that the progress of the cosmos towards it can only be like that of a child's hoop, or a Space Agency moon-shot. For it will be subject to chance off-course deviations and will have to be set on-target again by the equivalent of taps from the hoop-stick or impulses from guidance rockets.

There are conceptually two possibilities here. Either the guidance interventions are below the limit of observation set by the Uncertainty Principle and so cannot, in principle, be detected; or they are above it, in which case they will be detectable. Now the first possibility means that God himself cannot be subject to the Uncertainty Principle, or he would not know how and when to intervene. Consequently the notion of chance (as involving incomplete knowledge) would not apply to him, and Dr. Habgood's position (in this case) loses its foothold. The second possibility means that the progress of the cosmos will actually reveal discontinuities, just as the motion of the hoop or the moon-shot does, (but, unlike them, physically inexplicable). The discontinuities, unless large, might be very difficult to detect,[4] so the lack of recognized physical evidence for them does not tell conclusively against the suggestion; but there are few attractions for thought in the idea that the Creator, acting in effect from outside (since he normally leaves the outcome to chance) keeps the cosmos on target by innumerable interventions, too small to be physically obvious. This view is almost deistical, yet it seems that it is what the idea in question amounts to. But the strongest objection to it is that it is quite unbiblical. Bartholomew agrees that MacKay's view (that God's will is determinative even of single elementary events)[5] has 'logical force and sound biblical foundation'.[6] He misfires, however, when he attributes to MacKay the opinion that 'its logical possibility . . . is sufficient to establish its credibility'. Logical possibility is indeed necessary for credibility; but what MacKay (and the present author) have primarily emphasized is the biblical witness. How God rules his creation on the physical level is something surely we can expect to know only as he himself is pleased to tell us, and that is why the Bible's witness is all-important. Such statements as that in Psalm 104:24 ('O Lord, how manifold are thy works! In *wisdom* has thou made them all; the earth is full of thy creatures') and John 1:3 ('all things were made through him, and without him was not anything made that was made') hardly suggest a God who makes things by chance, i.e. who allows chance to take its course and throw up what novelties it will.

I made a suggestion earlier (Chapter 13) that at critical points in what the biologist calls evolutionary history mutational changes were not random (in his sense) but positively directed by the Creator towards the development of radically new forms of life. This suggestion has not the semi-deistical character of the view put forward by Dr. Habgood (as it may seem to have at first sight) since it regards *all* mutations (biologically random and biologically non-random) as equally God-given.

APPENDIX NINE

The Problem of Evil

The doctrine of the God as Creator sooner or later encounters the demand of theodicy: How can God's goodness be vindicated in a world knowing sin and suffering? This is not the place to attempt such a vindication; in any adequate sense only God himself can do this, and one day, no doubt, he will. But a few remarks may be appropriate.

Secularism does not escape the problem. It only changes its thrust and makes it less easy to define. In place of a God who can be challenged and questioned[1] secularism offers us only a shapeless, nameless, impersonal silence with which we must wrestle in the dark, uncomprehending and without hope. Scripture recognizes but does not resolve the problem; however, it makes it intellectually bearable and experimentally even something to be rejoiced in.[2] It is the former aspect we must briefly consider.

The interpretation we have given of the primal creation means that the predatory habit, pain, fear and death were features of the animal world before the coming of man. They cannot be explained therefore as the consequences of man's sin; they were a transient stage rather in the full outworking of the divine plan. (In a similar way, darkness and formlessness had been a transient stage on the way to light and life.) The plan envisaged (and envisages) something inexpressibly glorious[3] as a consummation which would (and will) fully justify the suffering and evil.[4] But, we may still ask, *how* will this final outcome justify things?

For answer we may turn to the great prophecy of Isaiah 11:1–9. In the *palingenesia*,[5] when God makes all things new, it will be *the universal knowledge of God* which secures 'fullness of joy' and 'pleasures for evermore'[6] for all God's creatures. But the Bible elsewhere seems to imply that the sort of knowledge this means—we might call it a vivid personal understanding of the Divine love[7]—is impossible apart from the actual experience of suffering;[8] and that this conclusion is true in an absolute sense and not merely because of the accident of a fallen world. For how could God commend his love so powerfully to us if we had not been sinners? How could our Lord have manifested the 'greater love' to us if we had not been in mortal need?[9] The father in our Lord's parable[10] was able to show his immense love to the younger son who had been lost just because he had been lost; what avenues were available for him to show his love to the correct elder son in the same degree? None, so far as we can see. This line of thought, stemming from the suggestion that the primal creation contained elements of pain, may provide a clue as to why the Creator has allowed evil at all in his world. If knowing the divine love is the ultimate blessedness for the creature, and if the divine love can be known fully only through the experience of sin, suffering and forgiveness[11] (as the prodigal son came to know it), then we have here a theodicy which not only goes some way to explaining the mystery of evil, but which also justifies the appellation 'good' to a creation 'subjected to futility'. And this is very much to our present point.

This opens up an interesting consideration. The suggestion is often made that God in omnipotence created man with the power to disobey, because only so could man yield the obedience of that love which the New Testament calls *agapē,* the love which chooses its object as esteemed and precious. Man chose to disobey, the Bible tells us, esteeming self-determination more precious than friendship with his Maker. God redeemed man at great cost to himself,[12] the Bible goes on to say, and will bring him ultimately to glory.[13] This raises another question for our consideration: what will prevent it happening all over again? Man in glory will presumably still possess power to disobey, else how could he *then* exercise *agapē*? What will defend him from doing again what man in Eden did? After all, some of the angels fell! The only answer the Bible seems to suggest is that it will be the remembrance of the redeeming 'love of God in Christ Jesus our Lord' that forever fixes his will in the right direction.[14] But this answer has very far-reaching implications.

It seems to mean that the settled loyalty throughout eternity of the human citizens of God's heavenly kingdom is *necessarily dependent on prior sin and suffering.* The rational creatures who comprise it will have power to disobey, as Adam did, but will never do so because, unlike Adam, they see in the midst of all 'a Lamb standing, as though it had been slain',[15] the ever-present reminder of 'the breadth and length and height and depth' of the love of the Creator for the souls he has made, a love holy and righteous too.[16]

The Bible, to conclude, does not give us an overall picture of an ideally-perfect creation that unfortunately went wrong and had to be redeemed as an originally-unintended consequence. Suffering and redemption, infinitely costly, were part of the plan from the beginning for both Creator and creature.[17] *Only so* could everlasting blessedness be secured, because *only so* could the divine love be made fully known. For our present study it is worth repeating that the important point is that this theodicy goes far towards justifying the verdict, 'God saw . . . that it was very good', for a primal creation that contained predation, pain and death. These words are a great affirmation of the worthwhileness of it all, whatever the appearance.

APPENDIX TEN

Extracts from Secularist Writers

Sir Julian Huxley in *Evolution, the Modern Synthesis* (my italics):

> Finally, mutations, while they seem to occur more readily in some directions than in others, *can be legitimately said to be random* with regard to evolution. . . . the directions of the changes produced by them *appear to be unrelated* either to the direction of the evolutionary change . . . or to the adaptive or functional needs of the organism. Evolutionary direction has to be imposed on *random mutation* . . . by [natural] selection.[1]

This is hardly an adequate discussion in this major work of a point of such cardinal significance!

> The ordinary [*sic*] man, or at least the ordinary poet philosopher and theologian, is always asking himself what is the purpose of human life . . . some . . . point to evolution as manifesting such a purpose.
> I believe this reasoning to be wholly false. The purpose manifested in evolution . . . is only an apparent purpose. It is as much a product of blind forces as is the falling of a stone to earth or the ebb and flow of the tides.[2]

Such extraordinarily blinkered statements are, alas, very common in the writings of leading secular theorists. But at least Huxley has the good sense to recognize it as *belief,* an article of his secular faith. The last sentence should be compared with the recent opinion of the cosmologist Sir Fred Hoyle, quoted earlier (Chapter 10).

George Gaylord Simpson in *The Meaning of Evolution*:

> We do know one negative fact: the results of mutations do not tend to correspond at all closely with the needs or opportunities of the mutating organism. It is a rather astonishing observation that the supply of this basic material for evolution *seems* to have no particular relationship to the demand. This accounts for much of the opportunism in evolution, and the nature of that opportunism in turn attests the random nature of mutation.[3] (my italics)

What is astonishing is that a scientific writer of Simpson's stature can base his whole philosophy of life on such a *seems.* Does no father ever make dispositions for his children which (to them) seem 'not to

correspond at all closely with their needs or opportunities', and yet
are both purposeful and wise? Where is Simpson's logic? He writes
later of

> . . . evidence . . . thoroughly conclusive evidence as I believe . . .
> that organic evolution is a process entirely naturalistic in its
> operation . . . Life is materialistic in nature . . . man arose as a
> result of the operation of organic evolution, and his being and
> activities are also naturalistic . . . purpose and plan are not
> characteristic of organic evolution and are not a key to any of its
> operations . . . the Universe . . . displays convincing evidence of
> their absence.
>
> Man was certainly not the goal of evolution, which evidently had
> no goal. He was not planned, in an operation wholly planless.
>
> Man is the result of a purposeless and natural process that did
> not have him in mind. He was not planned. He is a state of matter,
> a form of life, a sort of animal . . .[4]

This is all expressed very intemperately, and 'fundamentalist'
reactions to it in similar vein are understandable. It is acceptable only
as an expression of Simpson's own personal faith, without any
specifically scientific authority. Many will prefer the intuitions of the
'ordinary' man, as Huxley called him.

Ernest Mayr, *Evolution*:

> Man's world view today is dominated by the knowledge [*sic*] that
> the Universe, the stars, the earth and all living things have evolved
> through a long history that was not preordained or programmed
> . . .[5]

Prof. Mayr is another founder of the currently orthodox school of
Neo-Darwinism. His denial of divine purpose and providence is quite
outside his competence *as a scientist*. He is blundering badly when he
calls it knowledge, for its credentials derive ultimately only from his
mythology of chance.

Jacques Monod in *Chance and Necessity*:

> We say that these events [mutations] are accidental, due to chance.
> And since they constitute the *only* possible source of modifications
> in the genetic text, itself the *sole* repository of the organism's
> hereditary structure, it necessarily follows that chance *alone* is at
> the source of every innovation, of all creation in the biosphere.

Pure chance, absolutely free but blind, at the very root of the stupendous edifice of evolution: this central concept of modern biology is no longer one among other possible or even conceivable hypotheses. It is today the *sole* conceivable hypothesis, the only one compatible with observed and tested fact. And nothing warrants the supposition (or the hope) that conceptions about this should, or ever could, be revised. (Monod's italics)

Even today a good many distinguished minds seem unable to accept or even to understand that *from a source of noise natural selection could quite unaided have drawn all the music of the biosphere.* (my italics)

The ancient covenant is in pieces; man at last knows that he is alone in the unfeeling immensity of the universe, out of which he emerged only by chance. Neither his destiny nor his duty have been written down. The kingdom above or the darkness below: it is for him to choose.

It is almost incredible that a Nobel prizewinner should have written such confident but foolish and melodramatic claptrap. Along with the other extracts reproduced above, it should be remembered when judgement is being passed on what has emanated from sources pejoratively referred to as 'fundamentalist'.

Far more worthy of respect is the cautious approach of another Nobel prizewinner John Kendrew (also a molecular biophysicist):[6]

It is often said that scientists do not believe in miracles. And it is asked how the scientific approach can possibly 'explain' living organisms, especially man, who is the crowning miracle of the universe.

I think this formulation rather misses the point, for in fact biologists approach the challenge of complicated phenomena simply by saying, 'Let us see if we can discover how this animal works. Never mind how far we can go along the road of explanation, it is simply interesting to go along with it as far as we can, using whatever tools we can devise.' We do not even know if there *are* any limits.

Whether eventually there are limits beyond which one cannot go is a matter which the biologist, *qua* biologist, really does not care about, simply because he has far too many interesting things to do, right now, without worrying too much if he is going to come up against a dead stop later on.

At least Kendrew accepts the possibility that the methodology of science may come up against a dead-end. It is entirely reasonable to believe that it will do so. But even that is not a positive requirement for Creation! (see Chapter 14).

For a courteous but devastating criticism of Monod see D. M. Mac-Kay's Riddell Memorial Lectures *Science, Chance and Providence* (OUP, 1978) and Mary Midgley's *Evolution as a Religion* (Methuen, 1985).

APPENDIX ELEVEN

The Problem of DNA

The problem of the origin of DNA has recently been attacked in a novel (and promising) way by A. G. Cairns Smith.[1] The question of its origin may be compared to a more familiar situation—that of precision machine tools. These need precision machine tools to make them—so how did the first ones originate? There is clearly an answer to this, and it is not difficult to see what it is. Rough tools were first hand-held. These were used to fashion better tools, and these still better ones, until eventually to-day's precision lathes and so on resulted. It was a question of 'low technology' giving rise to 'high technology'. Cairns-Smith argues plausibly that naked 'low-tec' genes in the form of minute clay crystals deposited in the porous structure of sandstones began the whole process. Growing from the nutriment supplied by saturated solutions of rock minerals, encoding information by means of crystal faults (such as twinning), subject to replication by fracture as they grew (in a way that maintained the information content in each fragment), and susceptible to a simple form of natural selection, these naked inorganic low-tec genes multiplied, evolved, and began to act as centres for the chemical combination of water and carbon dioxide under the influence of ultraviolet light. There seems nothing unreasonable in this suggestion. In so doing they prepared the way for their own replacement by the much better high-tec machinery based on DNA and proteins. It is an attractive hypothesis, but as the reader will appreciate has little bearing on the argument of this essay.

APPENDIX TWELVE

God and Particle Physics

Is God the author of the detailed outcome of even quantum physical events? This conclusion, to which (it seems to me) the teaching of the Bible points, poses an obvious philosophical problem. How can it be reconciled, it may be objected, with that other and surely more firmly grounded biblical doctrine, the accountability of men and women to God for their actions, with the Bible's summons to them to choose right and to reject evil?[1] This is a difficult question, not to be answered adequately in a short appendix. But one or two points may be made.

(1) The problem is not uniquely a consequence of extending God's authorship of events down to the ultimate physical level. A similar problem is revealed in such uncompromising New Testament affirmations as those of Acts 2:23 ('This Jesus, delivered up according to the definite plan and fore-knowledge of God, you crucified and killed . . .'); Acts 4:27,28 ('. . . there were gathered together . . . Herod . . . Pontius Pilate, with the Gentiles and peoples of Israel, to do whatever thy hand and thy plan had predestined to take place'); and Philippians 2:12,13 ('Work out your own salvation . . . for God is at work in you both to will and to work for his good pleasure'). The list might be extended by many references; an Old Testament example is Genesis 50:20 ('you meant . . . but God meant . . .').

(2) A very similar problem arises inescapably also in purely secular philosophy as a consequence of the mind-body relationship. It may be put like this: if there is a one-to-one correspondence between states of the mind and configurations of the brain, and the latter obeys physico-chemical laws, isn't our sense of free will somehow an illusion? (Of course, nobody *lives* by this conclusion, whatever they may say).

(3) The wave-particle paradox in Physics has alerted us to the fact that apparently irreconcilable views may both have to be taken on board as valid expressions of the truth; further, that it may be exceedingly difficult to see our way through the problem this poses.

(4) This leads to an analogy which may help to relieve our intellectual tension. Imagine a boy lying face-down on a flat heath and watching a road crossing the expanse in front of him. Two vehicles are approaching each other along the road; a crash is inevitable! But no, they pass without incident. Of course, the boy is not puzzled. However vivid the play of his imagination he knew all along that there was a dimension, not obvious from his lowly

position, in which the vehicles could be displaced and so pass without encounter. The knowledge of this dimension,[2] not then advertising itself, resolved the paradox: two bodies coinciding at the point of 'impact' without disaster. Simple as this analogy is, it would seem to open an intellectual door of hope. We can surely believe that the divine 'dimensionality'[3] is rich enough to allow the determinative will of God to co-exist in the same occasion with both the will of man and physical causation.[4]

(5) There is an important consequence of the view taken here. Such biblical challenges as that expressed in Jer. 31:37 might seem to be in danger of being outflanked by advances in scientific knowledge. However, if God has chosen (as, for example, by the Uncertainty Principle) to set a limit to man's penetration into the workings of Nature, then this fact lays a permanent ground of validity for such challenges. This may be an illustration of one meaning of that remarkable verse of Scripture, Ecclesiastes 3:11 (see NIV, NEB). Impenetrable mystery will always remain.

(6) The biblical doctrine of Creation, however, hardly stands or falls with the view that God is responsible for the detailed outcome of every microphysical event. If chance had to be re-admitted at this fundamental level the argument of this essay would remain, in essentials, largely unaffected.

APPENDIX THIRTEEN

'God of Chance' by D. J. Bartholomew

My late friend Donald MacKay kindly gave me permission to use his review of this book by D. J. Bartholomew (SCM, 1984), referred to elsewhere in these pages. The review first appeared in *Religious Studies 21*, 622–4, 1986.

If the development of a satisfactory theology of chance required expert knowledge of statistics, then the author of this readable book, who holds a chair of Statistical and Mathematical Science at the London School of Economics, would be better qualified than most theologians who have tackled the subject. He is also a practising Christian who accepts fully the need to do justice to biblical data—a welcome if unfashionable emphasis in these days of do-it-yourself theologies for modern man.

Unfortunately the basic issues that need to be tackled—the sovereignty of God over natural events, and what is meant by calling him their creator—are not much illuminated by the technicalities of statistical theory; and they present numerous pitfalls for those who would cut theological corners in the interests of simplicity. For all Professor Bartholomew's evident awareness of the risks, I am not sure he has successfully insured against them.

In contending against Jacques Monod's atheistic argument in *Chance and Necessity* he begins well by insisting that 'to say that anything is caused by chance is, strictly speaking, a contradiction in terms. The chance hypothesis . . . is no more than an acknowledgement of our failure to find a cause. Tychism is not a rival to theism but is, rather, a state of agnosticism about causes' (p. 4). So far, so good. But later, and repeatedly, he forgets his own *caveat* with sadly confusing effects. 'Chance', he claims, 'was God's idea, and . . . he *uses* it to ensure the variety, resilience and freedom necessary to achieve his purposes' (p. 14). 'The universe was designed in such a way that *chance has a role* to play' (p. 30, my italics). Behind such claims there lies the recurrent refrain that 'chance is real' (pp. 94, 109), which seems to mean that even God himself finds chance events outwith his control.

This reveals what strikes me as the central theological weakness in Professor Bartholomew's case. If 'God' were merely the name of an invisible superhuman constructor-in-time and participant-in-time alongside us and the rest of his world, it would make sense to postulate (as Bartholomew does repeatedly), that, even for God, random processes were unpredictable and only partly controllable, although (like a human Monte Carlo computer programmer) he might use 'random strategies' (p. 98) to achieve his ends. But the God of trinitarian theism is not merely a 'constructor-participant' of and in our space time. He is also its *Author*. Bartholomew quite fails to grapple with the implications of this (equally biblical) emphasis, or indeed with the need for more than one 'personal projection' of the Divine Being. His argument conceives God in the image of a single person—a kind of 'superman'—who is thought of as active only in time (as we are) and omniscient, if at all, only as a predictor-in-time.

A trinitarian theist can accept that the 'superman' image captures much of what the biblical writers testify about God as an Agent in our space-time. There is biblical evidence of the view that the Divine Person who 'pleads with us' and 'yearns over us' in dialogue is not the same Person (though the same God) as the One who ordains all events 'according to his determinate counsel and foreknowledge' and who views our whole space-time *sub specie aeternitatis*. If so, it entails no self-contradiction to deny to the one the kind of explicit 'foreknowledge' attributed to the other; Bartholomew's argument would be much less objectionable as a sketch of the kinds of knowledge and control predicable of the Divine Person who meets us in our daily life, rather than of the Creator-Person-in-eternity.

But if we attach anything like the weight given in both Old and New Testaments to the concept of God as the eternal Author of our whole space-time, then to deny that 'random' events are in some sense ordained by Him would seem as incoherent as to deny that the human author of a novel 'ordains' all the events, both random and otherwise, in his (imaginary) creation. To say this is not at all, as Bartholomew suggests, to 'resolve the God or chance dilemma by abolishing chance' (p. 24). In the case of a human creation such as a novel it would make no sense to argue whether the toss of a coin (or the disintegration of a radioactive atom) in the novelist's created world was 'truly random' or 'ordained by the novelist'. It may perfectly well be *both*. The novelist is equally sovereign whether he creates a world in which every event has causal links with earlier events, or one in which some events occur without assignable causal precursors and so are correctly described as 'pure chance'. To avoid any 'novelist-or-chance dilemma' we have no need, not indeed have we the power, to 'abolish chance' from his creation; for the dilemma is illusory.

In dealing with awkward biblical passages such as Proverbs 16:33 that appear to teach God's sovereignty over 'random' events, Bartholomew objects to the idea that 'every event, however trivial, must express a particular intention of God' (p. 108). This he takes to be the belief behind the practice of sortilege, of which he is generally critical for good reason. The doctrine of divine sovereignty over 'chance' events, however, says nothing to warrant the idea that all or indeed any such events are intended to offer divine guidance to men. Even a human artist may sovereignly create a random distribution of objects or events without 'expressing a particular intention' by each one individually. There is nothing to prevent a sovereign God from guiding men occasionally if He will, through the casting of lots; but the idea that *every* such chance event has a communicative significance is not Calvinist (or any other) orthodoxy but superstition.

When discussing human freedom Bartholomew presupposes (as many do) that 'a degree of indeterminacy in nature is essential if human choices are not to be illusory' (p. 4). He seems unaware of the objection that although physical randomness in the brain might secure the *unpredictability* of a choice, it would seem to weaken rather than strengthen the chooser's responsibility for the outcome.

As with Pollard's earlier *Chance and Providence,* one cannot help warming to the devout and humble spirit of this book. As an antidote to the brashness of Monod's *Chance and Necessity* it is exemplary. Whether the theism it defends is quite that of biblical Christianity I am not so sure.

D. M. MacKay

Some Books Cited in the Notes

Bartholomew, D. J., *God of Chance* (London, SCM, 1984).

Blocher, H., *In the Beginning* (Leicester, IVP, 1984).

Cassuto, U., *A Commentary on the Book of Genesis* (Jerusalem, Magnes Press, 1961).

Clark, R. W., *The Survival of Charles Darwin* (London, Weidenfeld & Nicolson, 1985).

Cranfield, E., *Romans* (Edinburgh, T & T Clark, 1975).

Davies, P. W., *The Accidental Universe* (Cambridge, CUP, 1982).

Habgood, J., *A Working Faith* (London, Darton, Longman & Todd, 1980).

Hayward, A., *Creation and Evolution* (London, SPCK, 1985).

Hick, J., *Evil and the God of Love* (London, SCM, 1966).

Kidner, D., *Genesis* (London, Tyndale Press, 1967).

Mackay, D. M., *The Clockwork Image* (Leicester, IVP, 1974).

Mackay, D. M., *Science, Chance and Providence* (Oxford, OUP, 1978).

Monod, J., *Chance and Necessity* (London, Collins, 1972).

Pollard, W. G., *Chance & Providence* (London, Faber & Faber, 1959).

Taylor, G. Rattray, *The Great Evolution Mystery* (London, Secker & Warburg, 1983).

Von Rad, G., *Genesis* (London, SCM, 1961).

Westermann, C., *Genesis 1–11* (London, SPCK, 1984).

Notes

NOTES TO CHAPTER 1 (pages 11 to 16)

1 I was ordained in the Anglican non-stipendiary ministry five years before I retired from my chair in the University of London.

2 The Coronation Service, cf. Rom. 3:2; No. 20 of the Anglican 'Thirty-nine Articles'.

3 2 Tim. 3:16.

4 For the first, lest the argument should seem unacceptably circular, see Appendix 1; for the second see J. W. Wenham, *Christ and the Bible* (Tyndale Press, 1972, reprinted 1984).

5 e.g. the Jewish scholar U. Cassuto. See the reference in the Preface of his *Commentary on the Book of Genesis* (Jerusalem, 1961).

6 Acts 2:23; 4:28 (NEB); 2 Cor. 5:18f; John 18:11; Luke 24:25f, 44f.

7 On biblical premises, that is. The God of whom the Bible speaks is just as much Lord of the ordinary as of the extraordinary, of the natural as of the supernatural. cf. John 4:48; Matt. 6:11 with John 6:14; 1 Sam. 14:6.

8 On Joseph Smith's testimony it was written on gold plates whose whereabouts were revealed to him by a messenger from heaven. After he had translated them (a task for which he was miraculously equipped) they were taken away by the same heavenly being.

9 In a parallel fashion we may successfully offer purely naturalistic explanations for such general phenomena as the form and function of flowers, or the onset of rain; or for such special historical events as the selling of Joseph into slavery, or the retreat of the waters of the Red Sea. Without denying the validity of such explanations (in fact sometimes positively affirming them) the Bible insists that here is displayed the personal activity of God (Matt. 6:30; 5:45; Gen. 45:5; 50:20; Exod. 14:21). See also on the Origin of Life (Chapter 11).

10 1 Cor. 2:2,8; 15:3,4,23,51f.

11 The public return of Jesus Christ (Matt. 24:3,27,34).

12 At least, many of them are. See the comments by John Habgood written when he was Bishop of Durham (*op. cit.*, p. 50). His disapproval of the scientist's expectation of 'clarity, objectivity and definiteness' in the 'murkier and more humanly complex waters of religion and ethics' is surprising in one who believes in the Light of the World (cf. Luke 11:34–36; John 8:12).

13 This is an *a fortiori* argument about the ways of God of a type sanctioned by our Lord (Matt. 6:30; Luke 11:5–13; 18:1–7), as well as often used in the Old Testament (e.g. Exod. 4:11; Ps. 94:9).

14 And, to be topical, by the present (1987) Bishop of Durham, Dr. Jenkins.

15 cf. Deut. 30:11–15. There is a strange idea abroad that the very desire for an authoritative guide is a denial of mature spirituality rather than a consequence of it. The logic of this idea is never explained. It seems to have affinities with the Tempter's suggestion to Eve (see Appendix 2).

16 Our Lord's words in Matt. 11:25,26 and Luke 10:21, spoken with peculiar emphasis, are very much to the point here. So are Paul's in 2 Tim. 3:15 and 1 Cor. 1:26–29, and the Psalmist's in Ps. 119:99.

17 This formula is virtually synonymous with 'it is written' or the 'Scripture has said'. See Matt. 4:4,7,10; 9:13; 15:4; 19:4,5; 21:13,42; Mark 12:10,24; 14:21,27; Luke 4:17,21; 7:27; 21:22; 22:37; 24:27,44,46; John 5:39; 6:45; 7:38; 15:25.
18 Quoted by A. R. Peacocke, in his Bampton lectures, *Creation and the World of Science* (Clarendon Press, Oxford, 1979).
19 Science has traditionally been an authoritarian discipline. No scientist is a free thinker; cf. T. H. Huxley, 'Sit down before fact as a little child . . . follow humbly wherever . . . Nature leads, or you shall learn nothing' (Letter to Charles Kingsley, 23 September 1860); or Sir Cyril Hinshelwood's Preface to *Chemical Kinetics of the Bacterial Cell* (1946)—'it is better to be put in one's place by [Nature] than by any other authority'. Hinshelwood was President both of the Royal Society and of the Classical Society.
20 Thus Classical Physics was dethroned and Relativity Theory put in its place, not as homage to fashion, but because of a more accurate reading of Nature.
21 So the wave and particle aspects of matter and radiation have to be regarded as somehow reconcilable.
22 e.g. Reflections, mirages, diffraction patterns, are not what they seem to be at 'first sight'; nor is a sonic boom!
23 This does not necessarily imply in Science a uniformly upright moral character! Honesty is too obviously the only policy promising success. In the spiritual sphere the requirement is far more demanding.
24 e.g. cooking and keeping warm.
25 e.g. how to harness the electron.

NOTES TO CHAPTER 2 (pages 17 to 24)

1 Jacques Monod, *Chance and Necessity*. (Collins, London, 1972).
2 Gen. 28:16,17; Pss. 90; 139:7; Prov. 15:3; Zech. 4:10.
3 Deut. 4:12,15,16.
4 This is the least we can say. The Bible reveals God as tripersonal, three Persons in One God.
5 1 Pet. 4:19.
6 John 3:8; 1:18; Job 23:8,9.
7 Isaiah 31:3.
8 John 4:20–24.
9 Psalm 90:1.
10 See how the writer of Psalm 139 (e.g.) expresses the ideas of God's omniscience (vv. 2,3,4), omnipresence (vv. 8,9,10-), omnipotence (vv. 13,14, 15), and moral purity (vv. 19,20,23). All are in highly concrete terms.
11 Psalm 102:25,26. When God decreed that the heavens *inter alia* the 'work of his hands', should be *mutable,* he established time as a feature of the world. Hebrew thought has no other way to express this. See also Heb. 1:2 which is literally 'through whom he made the ages'.
12 For the connection of these with God as Creator see Isaiah 40:25,26, and 42:5,6.
13 For the connection of these ideas see e.g. Lev. 20:24–27; 1 Sam. 2:2; Isaiah 40:25.
14 See e.g. Psalm 102:19,25–27; Isaiah 31:1,3; 57:15; Rom. 1:25; Rev. 4:8–11.
15 Isaiah 28:17; Amos 5:24; 7:7,8; Luke 1:6.

16 'Covenant' is an extremely important biblical idea. See Exodus 34:1,6,7; Pss. 7:9,11,17; 103:17,18; Dan. 9:4,14.

17 Job 28:25–28 (NIV). See also Ps. 135:3,6,7; Matt. 5:45,48. The important idea of covenant applies to both: see Deut. 5:2,6f; Jer. 33:25.

18 This teaching is integral to the whole Bible: e.g. Gen. 8:21,22; Pss. 119:89–91; 146:5,6; Jer. 31:35,36; Matt. 5:45.

19 See R. Hooykaas, *Religion and the Rise of Modern Science* (Scottish Academic Press, 1972).

20 Rom. 1:17; 8:18–25.

21 For a discussion of the contradiction implicit in the idea of infinite Euclidean space see S. L. Jaki, *Angels, Apes and Men* (Sherwood Sugden, Illinois, 1983) pp. 83, 84.

22 Just as a 2D surface needs 3D space if it is to be curved into a spherical shape (and so lose its edges and become 'finite but unbounded').

23 *Mysticism and Logic* (Penguin Books, 1914).

24 'Time like an ever-rolling stream' (Isaac Watts).

25 'Footprints on the sands of Time' (Longfellow).

26 'Time flies'.

27 *Confessions,* translated Pine-Coffin, xi. 13,14; written about AD 400. (Penguin Classics, 1961).

28 *Ibid.,* xi. 5.

29 Isaiah 57:15; 63:15; 1 Tim. 6:16.

30 Pss. 123:1; 102:27.

31 For Newton, of course, they were not a *single* continuum.

32 The interesting consquences of this idea were worked out by A. T. Schofield in a book long out of print—*Another World: or the Fourth Dimension* (George Allen and Unwin, 1920). C. S. Lewis has some suggestive alternative comments in his section 'Time and Beyond Time' in *Mere Christianity* (Fontana Books, 1958).

33 John 10:36.

34 Isaiah 63:15 cf. John 17:5.

35 Compare the 'sending forth of the Spirit', creatively, in Ps. 104:30.

36 John 16:28; Gal. 4:4.

37 Augustine, *The City of God,* x.6. Healy's translation.

38 Job 10:5.

39 Creation out of nothing: see Chapter 4.

40 Acts 17:25; Rom. 11:36 (NEB).

41 J. B. Phillips.

NOTES TO CHAPTER 3 (pages 25 to 33)

1 Luke 20:4 (NIV).

2 Prof. James Barr, himself a strong opponent of the view here advocated, nevertheless writes, 'There is no doubt that Jesus, as depicted in the Gospels, accepted the ancient Jewish Scriptures as the word of God and authoritative'. *Escaping from Fundamentalism* (SCM Press, 1984), p. 18.

3 I believe I owe this metaphor to Dr. J. I. Packer. Its justification is superabundant in the Bible: see for instance Exod. 20:1; Ps. 19; Isaiah 1; 40; 55 etc.; Jer. 2 to 6; 7:25,26; Rom. 1:1,2; 9:25; 10:11; Gal. 3:8; Heb. 1:1,2; Rev. 22:16,18 cf. Deut. 12:32; Prov. 30:5,6. 'What Scripture says, God says'—Augustine (cf. Rom. 15:4,5; Rom. 11:32; Gal. 3:22).

4 We are thinking of the wave-particle complementarity of matter and

radiation.
5 Isaiah 53:1; Mark 14:61–64; Luke 20:9–19; John 3:11,32; 10:31ff.
6 Isaiah 66:2; Luke 10:21.
7 Ps. 2; Mk. 10:45; Luke 24:20,25–27, 44–47; Acts 2:23; 4:24–28; Rom. 8:28–32, 11:33–36; Gal. 4:4; Eph. 1:9–11; 1 Pet. 1:18–20. See further Appendix 1(iii).
8 Luke 10:21–24; 1 Cor. 1:18–24; 2:6–13.
9 Rom. 3:2. It is a consequence of the principle I am defending that while it may be an *important* truth that (e.g.) the prologue to John's Gospel is the result of John's mature theological reflection, the *all-important* truth is that it is God-given Revelation, God speaking by the Holy Spirit.
10 See D. Kidner, *Genesis: An Introduction and Commentary* (Tyndale Press, 1967), Henri Blocher, *In the Beginning* (Inter-Varsity Press, 1984).
11 Luke 20:9–19.
12 Kidner has drawn attention to this (*op. cit.*, p. 66). Other examples are 2 Sam. 12:1–6; Isaiah 5:1–6; Ezek. 16:17; Dan. 2:31–45.
13 Chapter 7.
14 Except in C. S. Lewis's sense—see Preface to *God in the Dock* (Collins, 1979).
15 e.g. Gen. 18:1f, 33; Exod. 3:1–6.
16 In his *sitz-in-leben* as the Germans say.
17 Matt. 1:22.
18 See for instance, Dan. 12:8,9; Luke 10:23,24; 24:25; 20:38 with 2 Tim. 1:10; 1 Cor. 10:11; Eph. 3:5, 6 with Isaiah 49:6; 1 Pet. 1:10, 11.
19 See Exod. 20:8–11; 31:12–17. I am far from suggesting that there were no thinkers in Israel who could have appreciated there and then the present suggestions about the 'days'.
20 *Screwtape Letters* XXIII (Collins, 1942).
21 Rom. 2:14, 15.
22 Rom. 1:20; cf. Isaiah 40:21.
23 U. Cassuto, *Genesis* (Jerusalem, 1961).
24 That these are mentioned in Isaiah 27:1 and elsewhere no more accords supernatural reality to them than does their mention in Exod. 12:12 accord it to the gods of Egypt, or for that matter, that my own comment that 'God is the Master of Chance' is meant to accord it to the latter.
25 cf. Deut. 4:19; Jer. 8:1, 2; cf. von Rad, *loc. cit.*
26 see Appendix 1 (ii).
27 John 18:11; Acts 2:23; 2 Cor. 5:19.
28 Any more than it would be adequate to understand Paul's Epistle to the Romans *merely* as apostolic doctrine.
29 *Commentary on Genesis,* English trans. (SCM, London, 1961). (Von Rad died in 1971.)
30 Isaiah 66:2.
31 These refer to the Documentary Hypothesis (of the origin of the Pentateuch) associated especially with the names of Graf and Wellhausen.
32 Isaiah 24:19, 20 (NEB); Jer. 6:10c; 8:9.
33 In this connection I cannot recommend too strongly the superlative exposition of Prof. Henri Blocher, *In the Beginning* (IVP, 1984).

NOTES TO CHAPTER 4 (pages 34 to 42)

1 It is sometimes a simple synonym for 'create' (see, for example, Gen. 5:1).

2 See, for instance, Numb. 16:30; Pss. 51:10; 104:30; Isaiah 48:6,7; 65:17; 31:22; Eph. 2:15; 4:24; Col. 3:10; cf. 2 Cor. 5:17.

3 The dress designer and the hair stylist know how to cash in on this!

4 Von Rad: 'It is correct to say that the verb *bārā'*, 'create', contains the idea both of complete effortlessness and *creatio ex nihilo*, since it is never connected with any statement of the material' (*Genesis*, p. 49).

5 Ps. 33:6 (NIV).

6 John 1:1–3.

7 Something that is, about what is often called (misleadingly) 'secondary creation'. Creation in the Bible is never 'secondary'; it is the unanalyzed movement from the mind of God into physical existence in our world.

8 Gen. 2:7; 2:22 (NEB).

9 Jer. 1:5; Job 31:15; Isaiah 44:2,24.

10 Ps. 139:13,15—'form', 'knit together', 'made', 'intricately wrought' (RSV).

11 The nearest the Bible comes to breaking this rule is in Ezek. 21:30 but this is quite untypical and can hardly affect our conclusion.

12 This statement neglects, for simplicity, the world of created *spirits*.

13 Rev. 3:14 (RV); cf. John 1:1,3 (NEB); Col. 1:15,16 (NEB).

14 Heb. 11:3.

15 See the note on 2 Pet. 3:5–7 in Appendix 3.

16 Isaiah 48:6,7,13,20; Ps. 51:10. 'Just as ch. 1 understands nature as created by God's word, so the Old Testament knows history also as created by God's word. See Isaiah 9:7; 55.10ff . . .'. von Rad, *Genesis* (p. 52).

17 Eccles. 12:1. See also Isaiah 43:15; Mal. 2:10; 1 Pet. 4:19.

18 A. F. Kirkpatrick, *The Psalms, Cambridge Bible for Schools and Colleges* (Cambridge, 1906). The reference is to Ps. 104:29,30.

19 This analogy throws a little light on the problem of sin, for as Dorothy Sayers noted, (*The Mind of the Maker*) fictional characters, once created, do have a sort of free-will of their own. They can't just be made to do anything the author wishes, or they lose credibility. See also the television analogy below.

20 Of course, stories by human authors do have characters dropping in suddenly—but that is because our stories are only narrow slices of the whole. On the other hand, as we shall see below, God is under no *necessity* to link every event in his story to earlier events in a law-like way; he can institute what we call *miracle*.

21 Quantum Physics has recognized the same principle.

22 1 Tim. 6:16.

23 Isaiah 6:3; 57:15.

24 Ps. 19:1–5; Isaiah 40:26; 46:3–11; Ps. 119:89–91; Eph. 1:9–11; Rev. 10:6,7.

25 Isaiah 40:25,28; 44:20; Rom. 1:25.

26 Pss. 115:3; 135:6; Isaiah 40:13,14,26,28; Rev. 4:11; 2 Tim. 2:13; 1 Cor. 8:6 (NIV, NEB).

27 So we cannot 'explain away' the existence of predators, poisonous snakes, or noxious insects as if they were not his creatures (see Ps. 104:21; Numb. 21:6; Exod. 23:28; John 1:3).

28 Pss. 89:11; 93; 103:19–22; Dan. 4:35; Luke 8:25; Rev. 19:11.

29 Acts 7:51; John 19:11; Rom. 14:10–12; Rev. 20:11,12.

30 Deism was a system of natural (as opposed to revealed) religion which began in England in the late 17th century. It became very influential in France and Germany. Some of its great names are Toland, Tindal, Rousseau, and Lessing. The account of Deism here given is, of course, very simplified.

31 For the first see Ps. 90:1; Acts 17:28; Col. 1:16; for the second Job 33:4; Ps. 139:7–10; Jer. 23:24; Eph. 4:6.
32 Col. 1:16,17.
33 J. B. Lightfoot, *St Paul's Epistle to the Colossians* (London, 1879).
34 Heb. 1:3.
35 Rev. 4:11 (NEB, NIV). The two verbs using 'create' (one active and one passive) are aorists, signifying a past action complete in itself; 'have their being' is the imperfect of 'to be', literally 'were being'.
36 Matt. 5:45; 6:26,30.
37 Pss. 104:10, 27; 135:7; 148:8; Jer. 31:35; Amos 4:7–10,13.
38 Dan. 5:23; Job 34:14,15.
39 See Appendix 3; and note the two tenses in Jer. 10:12,13.
40 D. M. MacKay, *The Clockwork Image* (Inter Varsity Press, 1974).
41 Compare Isaiah 51:6; 2 Pet. 3:11,12.
42 *op. cit.,* p. 58.
43 That it is God who upholds this principle is implied by Paul's remark in Gal. 6:7.
44 See, for instance, Gen. 8:21,22; Pss. 36:5,6; 119:89–91; Acts 14:15–17.
45 It is translated 'of itself', 'of its own accord'.
46 Jer. 14:22 (NEB); Joel 1:4–7; 2:1–11,25; Jonah 1:4,17.
47 Philosophically this view may be very difficult. But it is not uniquely so. Other precisely-defined views are equally difficult, or more so. The matter is further discussed in Chapter 11 and Appendix 12.
48 Deut. 4:15–19; Rom. 1:21–23; contra Neh. 9:6; Ps. 148.
49 Ps. 102:25–28; Matt. 24:35; 2 Pet. 3:10–13.
50 *Ever Since Darwin: collected essays,* Pelican, 1980.
51 Gen. 1:27,28; the verses are closely connected.
52 Ps. 145:9,16; Isaiah 40:11; see also John 10:11.
53 Luke 12:42–48; Deut. 25:4. In Gen. 1 man is charged *in general* to 'subdue' and 'have dominion'; in Gen. 2 where it comes to *particulars,* he is to 'serve' (*'ābad,* till, means this basically) and to 'guard', 'care for' (*šāmar*). (A. A. Hoekema, *Created in God's Image* Paternoster/Eerdmans, 1986).
54 It would be difficult to substantiate a claim that the teaching of man's origin by natural selection had ever advanced human welfare. On the other hand, the idea of the 'survival of the fittest' (or Darwin's 'preservation of favoured races in the struggle for life') has had many pernicious effects. This doesn't prove its falsity—it may be *true* though *unhelpful* that one has a criminal ancestry! But it does prove that it is something to be just accepted (if true) and not emphasized, particularly to the young. It is the *Bible's* teaching that should be underlined.
55 cf. S. Weinberg, *The First Three Minutes* (André Deutsch, London, 1977) 'The more the universe seems comprehensible, the more it also seems pointless'.
56 cf. C. S. Lewis, *The Pilgrim's Regress* (G. Bles, 1943).
57 *Mysticism and Logic* (Penguin Books, 1953). Russell more or less confirmed this attitude in the Preface, written in 1917.

NOTES TO CHAPTER 5 (pages 43 to 49)

1 Ps. 145:9,10; Ps. 148; Luke 12:6.

2 This is often referred to as the doctrine of the 'fixity of species'. The proponents of this view prefer the wider and less precise term 'kinds'. However, any distinction between *species* and *kinds* inevitably opens the door (some way at least) to an origin of species by natural processes.

3 D. Kidner, *op. cit.*

4 See further Chapter 10.

5 Gen. 1:20 RV [margin]; 1:24.

6 See further Appendix 3(ii).

7 The initials used refer to translations of the Bible, mostly modern, are as follows: RSV Revised Standard Version; RV Revised Version; NEB New English Bible; NIV New International Version; JB Jerusalem Bible.

8 I owe this expression to the zoologist Gordon E. Barnes.

9 Compare the case of Ezek. 47:10 above.

10 Gen. 1:24–27.

11 Gen. 2:7,19. This clearly expresses a *terminal* truth in the same way as Gen. 3:19b,c does. Neither is to be taken in a physically-direct sense; cf. Gen. 18:27; 1 Cor. 15:47f.

12 John 1:14,18.

13 Gal. 4:4; Rom 1:3,4.

14 Heb. 2:17.

15 Col. 2:9.

16 Gen. 3:8; Job 21:14; Rom. 1:18.

17 Isaiah 65:11f.

18 Isaiah 44:20.

19 Isaiah 41:23f.

20 Isaiah 57:13; Jer. 10:11.

21 Ps. 139:1–5; Isaiah 40:27,28; Jer. 23:24; Luke 12:6,7.

22 Prov. 16:33.

23 Josh. 18:10; Pss. 16:5,6 (NIV); 47:4; cf. Acts 1:24–26.

24 Gen. 49:10. Jersualem was, in fact, just inside the neighbouring small tribe of Benjamin, closely associated with Judah (1 Kings 11:32; 12:21).

25 1 Kings 22:34.

26 This story illustrates one conspicuous aspect of the Bible's teaching (complementary to others): God's activity is not to be thought of as something superadded to the common course of history, or injected into it. Rather, it fills the whole of history (see such suggestive passages as Gen. 45:4–8; Deut. 32:8; 1 Kings 11:14; Ps. 135:6–12; Amos 3:6; Matt. 26:31 with Zech. 13:7; Luke 22:37 with Is. 53:10,12; Acts 17:26; Rom. 9:17; Eph. 1:9–11; and *passim*. For further remarks see Appendix 12.

27 Gen. 37:25,28; cf. Gen. 15:13f.; 50:20.

28 Ruth 2:3; cf. Matt. 1:5.

29 Luke 2:1; cf. Micah 5:2.

30 Pss. 112:6; 121:3 where the Hebrew verb is the same as in Pss. 93:1, 104:5.

31 The cheetah has affinities with the dog family as well.

32 i.e. sub-atomic events (see further Appendixes 8 and 12).

33 John Habgood, *A Working Faith* (Darton, Longman and Todd, 1980). D. J. Bartholomew, *God of Chance* (SCM Press, 1984). On this book, see Appendix 8 and the review in Appendix 13.

34 J. R. Moore, *The Post-Darwinian Controversies* (Cambridge University Press, 1979).

NOTES TO CHAPTER 6 (pages 50 to 58)

1 cf. Isaiah 11:9.
2 Gen. 1:4,10,12,18,21,25.
3 The Hebrew verb *'āśâ* means equally both 'do' and 'make'. Compare Gen. 2:2 in RV and RSV; and Gen. 1:31 and 2:2 (where the verb is the same) in the NIV, RSV, NEB. I have adopted the translation 'done' here.
4 Gen. 1:31.
5 Gen. 9:2,3.
6 Isaiah 11:9.
7 Mark 10:18.
8 Exod. 3:8.
9 e.g. Matt. 13:39; 25:41.
10 1 Pet. 3:22.
11 Eph. 6:12.
12 The witness of the Bible; the sheer *concreteness* of evil; occult phenomena and present-day demon-possession; the experience of dread; primitive intuition. See the sensible remarks of C. E. B. Cranfield, *Cambridge Greek Testament Commentary on Mark,* (CUP, 1977), p. 75.
13 Attempts have been made to replace the idea of a personal Devil with impersonal philosphical conceptions such as the *Das Nichtige* of Karl Barth (cf. John Hick, *Evil and the God of Love,* SCM, 1966). The results are without practical power, spiritually and ethically. It is best to take the Bible in its plain and consistent sense, however great the problem for theodicy.
14 As Gen. 3 makes plain. For the identity of the tempter see Matt. 4:1; Mark 1:13; 2 Cor. 11:3; 2 Thess. 2:9,10; Rev. 12:9.
15 1 Cor. 15:24–26 cf. Matt. 25:31,41; Heb. 2:14; 1 John 3:8.
16 1 Cor. 15:24.
17 Heb. 2:14.
18 Rev. 12:7–12—'by the blood of the Lamb and by the word of their testimony' cf. also Luke 10:17,18; Rom. 16:19,20.
19 Eph. 1:10; 3:9,10,11; Col. 1:20.
20 Gen. 12:3 (NIV).
21 Dan. 9:24.
22 I shall limit my comments almost entirely to the living world, though the mandate undoubtedly covers also the non-living. It is, in fact, a mandate for science and technology of all sorts.
23 'The expressions for the exercise of this dominion are remarkably strong: *rādā,* "tread", "trample" (e.g. the wine press); similarly *kābaš,* "stamp"'. von Rad, *Genesis,* p. 60.
24 Joshua 17:8, 18:1 ('land' is the same word as 'earth' in Gen. 1:28 and 'subduing the land' means 'subduing the inhabitants'.).
25 Zech. 9:15 (literal, cf. RV); Mic. 7:19.
26 See Appendix 4(i) on the 'great sea monsters' of Gen. 1:21.
27 Even today some men and women have remarkable power to establish friendships between animals naturally enemies. Compare the comment about our Lord in Mark 1:13 and cf. Mark 11:2.
28 Matt. 21:16 (quoting from the Septuagint); Heb. 2:6–9.
29 Had man faithfully fulfilled his mandate the outcome, I suggest, would have been as described in Isaiah 11:6–9. The critical factor in securing this is given in verse 9—the universal knowledge of God. So far from disseminating this knowledge, Adam lost it himself.

30 *Commentary on Genesis* (see on 1:29,30 and 9:3). See also the fine recent commentary by Henri Blocher *op. cit.*, p. 209 note.

31 Gen. 4:2.

32 Gen. 1:24. The Hebrew word translated 'cattle' here can mean simply 'beasts', but the verse is usually interpreted as suggested (cf. NIV). The reference in Ps. 8:7 seems to put it beyond reasonable doubt that animals for domestication are meant.

33 Gen. 4:4 (NIV, NEB, GNB); cf. the Peace Offering, Lev. 3:1–5; 7:11–15.

34 Gen. 4:3. 'In the course of time', lit. 'at the end of days', seems to imply this.

35 Man's first clothing seems to have been of animal skins, not fabricated wool (cf. Gen. 3:21).

36 'Food' here is the general term *brōma* (cf. Matt. 14:15). It certainly commonly includes flesh (cf. Rom. 14:2,15,20, where it is used three times, with 14:21 where 'flesh' and 'wine' are mentioned explicitly as included in the term; cf. similarly 1 Cor. 8:13).

37 The aorist tense here indicates an action complete in itself. Had Paul wished to refer to the *continuing* creation of animals, generation by generation (Ps. 104:30) he would surely have used the present.

38 Heb. 7:16, cf. Rom. 6:9,10.

39 Luke 24:42; John 21:9,10.

40 Matt. 19:4,5. The phrase quoted is from the Marriage Service (BCP).

41 'Futility' (RSV); 'frustration' (NIV, NEB); 'inability to attain its purpose' (JB).

42 See such passages as Rom. 10:6–8; 11:34,35; 1 Cor. 1:20; 2 Cor. 3:18; 4:6; Gal. 1:15 etc. Note further how unambiguously Paul refers to the Fall in Rom. 5:12–21. There is no such plain reference here.

43 This conclusion is not altered if we refuse to regard the Fall as a matter of *history*; what is important is that Paul himself regarded it as such.

44 e.g. 1 Cor. 8:6; 11:12; Col. 1:16,17.

45 cf. 1 Cor. 2:7; Eph. 1:4,7; 2 Tim. 1:9,10 (NIV, NEB); 1 Pet. 1:18,20; Rev. 13:8. The significant phrases are 'before the foundation of the world', 'before times eternal', 'before the ages'.

46 Rom. 11:32; cf. 1 Cor. 1:18–21.

47 Contrast Isaiah 42:4; John 17:4.

48 Rom. 8:20,21.

49 See C. E. B. Cranfield, *op. cit.* on Rom. 8:20.

50 A dog suffers when its master takes to drink—that is an outcome of *solidarity*. But it doesn't undergo anatomical and physiological changes which alter its whole biology!

51 Luke 22:53b. cf. Col. 1:13. In both cases the key word is *exousia* which almost invariably in the N.T. stands for authority *vested in a person*. The Power was personal.

52 Such as predators and parasites.

53 cf. Isaiah 48:18,19.

54 Rom. 8:19,20.

NOTES TO CHAPTER 7 (pages 59 to 67)

1 Thus von Rad, *op. cit.*, p. 25: 'As regards the creative genius of the *Yahwist's narrative* there is only admiration. Someone has justly called the artistic mastery in this narrative one of the greatest accomplishments of all

times in the history of thought.' (The *Yahwist* is the name given by critical scholars to the presumed author of the story of Eden).
2 Deut. 29:29; Matt. 4:4; 9:13; Rom. 15:1–6; 1 Cor. 4:6; 2 Tim. 3:14–16.
3 We are thinking here of the myths of Creation and Origin. '. . . the primary function of the myth is to maintain the stability of the present state; it is this that is common to the whole vast circle of stories about the creation or origin of the world and of human beings.' C. Westermann, *Genesis 1–11*, (SPCK, 1984). '. . . the recital . . . has the power to establish and ensure the continuity of human life.' R. Pettazzoni, quoted by Westermann, *loc. cit.*
4 Rom. 5:14,15.
5 See, e.g. Luke 3:1; Gal. 4:4; Acts 7:17; Luke 24:39; Joshua 4:21f; etc.
6 Matt. 16:17.
7 Gen. 12:1; Exod. 3:1; 1 Cor. 2:8.
8 Luke 10:21–24.
9 Mark 7:24f, 32f; 8:22f; Luke 5:14; 8:51,56; contrast Rev. 13:13–15.
10 Matt. 6:26; John 6:14; 14:10.
11 John 4:48.
12 John 10:38; 14:11; contrast Matt. 16:4; John 6:30.
13 Job 2:10; Ps. 147:7–9, 15–18; Prov. 16:33; Rom. 8:28.
14 As, for instance, the death of Herod (Acts 12:23).
15 I am not sure to whom I owe this analogy.
16 cf. also Pss. 119:73; 148:5—God's *hands* form, his *word* creates.
17 Heb. *yāsar.*
18 Jer. 1:5; Isaiah 44:10. There is every reason to regard the language of Gen. 2:7 as metaphorical, as in Isaiah 45:9; 64:8; Jer. 18:6.
19 Gen. 2:8,15. Robert Ardrey (*African Genesis,* opening sentence) has been hasty here.
20 Gen. 2:9. See further below, Chapter 8.
21 cf. Isaiah 40:24; Jer. 2:21; 18:7,8; Ezek. 36:35,36.
22 Brevity is the soul of more than wit; cf. Prov. 9:16b,17; 30:20 for parallel instances.
23 Gen. 41:25; Isaiah 45:15; 2 Cor. 4:16–18; Heb. 11:27.
24 Matt. 5:45.
25 Prov. 16:33; Matt. 10:29 (RSV, NIV, NEB). Compare the Collects (BCP) for the 8th Sunday after Trinity and the 2nd after Epiphany; see also Eph. 1:11 and Rom. 8:28.
26 Matt. 10:30.
27 See, for instance, S. J. Gould's essay on 'Uniformity and Catastrophe' in *Ever since Darwin* (Penguin, 1980).
28 cf. Prov. 12:11 where the same word (*'ābad,* till) is used. Of course, it implies technology.
29 Isaiah 28:23–29.
30 See W. H. Thorpe, *Animal Nature and Human Nature* (Methuen, 1974).
31 Gen. 1:22,28.
32 Gen. 2:16,17.
33 *nepeš* is variously rendered 'creature', 'soul', 'being', 'person'. *nᵉšāmâ* ('breath') seems to be the endowment of mankind uniquely in the Bible. With one possible exception (Gen. 7:22) it is never used of the animals. Joshua 11:14 seems to confirm the distinction.
34 The significance given here to the body makes it theologically quite unacceptable to deny (as it is fashionable to do today) that our Lord's *body* had a share in the glory of the Resurrection.
35 Heb., literal.
36 Other reasons have been suggested for the deep sleep: aesthetic—it

sustains the beauty of the story (Cassuto, *op. cit.*); and theological—it would have been inappropriate for man to watch the Creator actually at work (von Rad, *op. cit.*).
37 Isaiah 29:8.
38 Gen. 28:12f; cf. the pregnant passage Gen. 15:12–21, where a similar 'deep sleep' (Gen. 2:21) is pervaded by a revelatory vision.
39 Daniel 2.
40 Deut. 8:3.
41 Matthew Henry, *A Commentary on the Holy Bible* (1710).

NOTES TO CHAPTER 8 (pages 68 to 75)

1 Gen. 2:17.
2 D. Kidner, *op. cit.*; G. von Rad, *op. cit.*; C. S. Lewis, *Voyage to Venus.* For the whole content of this Chapter see the excellent discussion in H. Blocher, *op. cit.*
3 Gen. 3:7–12; 4:8.
4 Luke 9:23; John 12:26; Rom. 6:3–11.
5 Rom. 5:12,15–21; 1 Cor. 15:21,22; Col. 3:1,3. Consider also the locale (a garden); the role of the Tempter (cf. Luke 22:53); the aloneness of the victim; the immediacy of the prospect, pleasure (Gen. 3:6)—or pain; and so on.
6 Of the countless instances see Gen. 16:10–13; Exod. 3:1–6; Num. 22:21–31; 2 Kings 6:17; Luke 1:26–31; 3:21,22; Acts 2:1–4; 9:1–9.
7 M. B. Foster, *Mystery and Philosophy* IV (SCM Press, 1957).
8 See as examples Augustine's *Confessions,* Book IIX; or at the other extreme the testimony of Fred Lemon, *Breakout* (Lakeland, 1977).
9 e.g. Cassuto, *op. cit.*
10 The Bible nowhere asserts that *all* temptation comes through Satan; cf. James 1:14.
11 John 13:27; Matt. 16:23; Luke 8:32.
12 The idea that it was the eating of an *apple* is due probably to a confusion between the Latin words *malus* (evil) and *malum* (apple). A metaphorical understanding of Eve's act is favoured by the description of our Lord's corresponding act of obedience (See Matt. 26:39; John 18:11). There are arguments on both sides. See further, Appendix 5(i).
13 Gen. 5:24; Heb. 11:5.
14 Gen. 2:9; 3:22; Ezek. 47:9,12; Rev. 22:2,14. As with the other tree, the language may be either literal (sacramental?) or metaphorical. For the latter case see the note below, p. 171 n.21.
15 e.g. by the onset of a flat electroencephalogram.
16 John 17:3. Death is correspondingly linked with ignorance, Eph. 2:1; 4:18. This connection is often noted in the Bible: see Pss. 6:5; 88:10–12; Eccles. 9:5,6,10; Ezek. 37:13,14.
17 Deut. 30:20; Job 3:20; Ps. 36:9; John 1:4; 8:12; 1 John 3:14.
18 The Bible *as a whole* is fundamentally about relationships. The very names Old 'Covenant' (as 'Testament' should be translated) and New 'Covenant' should make this clear.
19 Paul's treatment of the subject in Rom. 5:12ff and 1 Cor. 15:21,22 puts this beyond dispute for those for whom Scripture is 'God speaking'. cf. also 1 Cor. 15:56.
20 See e.g. Luke 9:60, 15:24; John 5:24,25; Rom. 6:13; 8:6; Eph. 2:1,5; 1 Tim. 5:6; 1 John 3:14.

21 The Tree of Life may itself stand metaphorically for the spiritual truth expressed in John 15:1–6; cf. also John 6:52–58; Col. 3:3,4; 1 John 5:12. Henri Blocher seems to take this view (*op. cit.* pp. 122–125).
22 Bertrand Russell, *A Free Man's Worship* (*op. cit.*).
23 Ps. 90:9.
24 Eph. 4:18; Col. 1:21.
25 Luke 15:13; cf. also the verdict of Rom. 1:18–32.
26 Anxiety and fear are thus often stressed in the Bible as outstanding consequences of a fallen nature (contrast Ps. 91:1–6).
27 *The History of Nature* (Routledge and Kegan Paul). His reference is to Gen. 3:7.
28 2 Sam. 13.
29 Gen. 3:16.
30 One is reminded of the increasing threat of the new disease AIDS.
31 Ruth 2:4; Ps. 104:31; Neh. 4:6; 12:27,43; John 4:32,34; 15:11; Eph. 6:5–8.
32 Accentuated no doubt by the change from the horticulturally-favoured environment of Eden to the tougher and more demanding regions outside.
33 Hence the need for such injunctions as Deut. 22:6; 25:4 cf. also 5:14.
34 Isaiah 24:4–6; Lev. 18:27,28.
35 *Contra* Lynn White, Science *155,* 1203, (1976).
36 Gen. 3:15.
37 Rom. 5:17.
38 See especially Rom. 5:12–21; 1 Cor. 15:20–50; John 8:44.
39 Mark 10:6–9. It is difficult to see how myth in the present sense can have any real authority on the point at issue here.
40 Such as the Exodus from Egypt, and the Babylonian Captivity and the Crucifixion under Pontius Pilate.
41 That plants grew spines, for instance, and animals became adapted as predators or parasites.
42 For instance, the curse does not mention the animals (except the serpent) yet the changes they suffered are held to have been prodigious.
43 See further, Appendix 7.

NOTES TO CHAPTER 9 (pages 76 to 80)

1 So E. J. Young, *Genesis 3* Banner of Truth Trust (1966). Cassuto, *op. cit.*, p. 170: 'It is not Adam's own reason, for in that case he should have said: "because she *shall be* the mother of all living" '. Cassuto regards the name, in the lips of Adam, as signifying *Female Serpent!*
2 Like the father of John the Baptist (Luke 1:13).
3 Verse 15; cf. Gal. 4:4 where the verb is, unusually, not 'born' (from *gennaō* as in Gal. 4:23) but literally 'came to be' (from *ginomai*).
4 So E. J. Young, *op. cit.*; D. Kidner, *op. cit.*; G. von Rad, *op. cit.*
5 Gen. 17:5,6 cf. 15:5,6.
6 cf. I. H. Marshall, *Acts,* Tyndale N.T. Commentaries (IVP, 1980).
7 cf. E. M. Blaiklock, *Acts,* Tyndale N.T. Commentaries (IVP, 1959).
8 For the meaning of 'solidarity' see Chapter 13.
9 Gen. 4:14.
10 Rom. 5:14 (RV).
11 cf. Gen. 5:4.
12 Gen. 4:17.
13 The nearest occurrences of the same word are Gen. 10:11,12; 11:4; 18:24.

14 D. Kidner, *op. cit.*, p. 29.
15 cf. Gen. 5:6,18,25 but see also 5:9,12,15.
16 H. Blocher, *op. cit.*, pp. 200–203.
17 e.g. Sir Ambrose Fleming, *The Origin of Mankind* (M.M.&S., 1935). See also the excellent discussion in D. Kidner, *op. cit.*, pp. 26–31.
18 C. E. B. Cranfield, *The Epistle to the Romans* (T.&T. Clark, Edinburgh, 1975).
19 Note the emphasis on 'man' in Rom. 5:12,15,17,19; 1 Cor. 15:21,45,47; Phil. 2:8; 1 Tim. 2:5; Heb. 2:14,16,17; 5:1; and our Lord's favourite self-designation, 'Son of Man'.
20 For Adam see Gen. 5:1–3. Adam the individual was given as personal name what was the designation also of the species (*adam*); he was thus appointed representative of the race. For Jesus Christ see Heb. 2:7,9; 5:1,4, 5,10 etc.
21 Gen. 2:8; 5:1,2; 12:1; 17:5; Isaiah 9:1,2; Matt. 2:1,22,23.
22 See Luke 3:8; John 8:39; Rom. 4:16,17; 9:6–8; Gal. 3:7,29; cf. Is. 14:1; 56:3–8.
23 As the stories of Caleb, Ruth, the Kenites and other non-Hebrews (see Gen. 17:12,13) show; for the church, see John 10.16, Eph. 3:6.

NOTES TO CHAPTER 10 (pages 81 to 88)

1 Or say, that the sum of two and three is five.
2 Such as, that parallel straight lines never meet.
3 Rev. 4:11.
4 Put differently, scientists believe that events are predictable on the basis of observed precedent, but not on the basis of *a priori* rational principles. Quantum physics goes even further; atomic events are quite unpredictable, except in terms of probabilities.
5 This seems a fair statement of Bertrand Russell's view; see *loc. cit.*, chap. 2. Much modern existentialist literature sees human life as basically absurd. See *Waiting for Godot* or *Endgame* by Samuel Beckett.
6 See, for instance, Sir Bernard Lovell, *In the Centre of Immensities* (Granada, 1980).
7 It also holds the planets in their orbits.
8 The electron and positron are the two most familiar examples of this.
9 There are other possible interpretations of the present matter/radiation ratio, but they are equally mysterious.
10 P. C. W. Davies, *The Accidental Universe* (CUP, Cambridge, 1982), p. 91. Davies is Professor of Theoretical Physics at the University of Newcastle-upon-Tyne. He does not write as a religious believer.
11 *Ibid.*, p. 73 (his italics).
12 *Ibid.*, p. 95.
13 *Ibid.*, p. 111.
14 Quoted by Davies, *ibid.*, p. 118.
15 Greek *anthrōpos,* man.
16 Other suggestions, such as that based on the 'Many Worlds' interpretation of the quantum theory, will tax the credulity of the generality no less. What price common sense?
17 Gary Zukav, *The Dancing Wu-Li Masters: an overview of the New Physics* (Fontana/Collins, 1979).

NOTES TO CHAPTER 11 (pages 89 to 96)

1 Prov. 16:33; Acts 1:24,26.

2 Following a suggestion of Prof. D. M. MacKay; see his lucid Riddell Memorial Lectures *Science, Chance and Providence* (Oxford Univ. Press, 1978).

3 'Probably' is not really a significant word here. We may assume they would. Final digits only are taken to ensure that the *a priori* probabilities are equal. The point is technical.

4 Or rather, on the peer status of the Telephone Authority and the would-be predictor.

5 'Natural selection operates upon the products of chance and knows no other nourishment.' J. Monod, *Chance and Necessity,* (English translation Collins, London, 1972).

6 I use the word 'creationist' (in quotes) for those who deny evolution as a providential mechanism.

7 That is, random by any test open to us to apply. I shall deal in a moment with the question of whether gene mutations *are* random.

8 1 Kings 22:34 (NIV, NEB, JB). (See above, Chapter 5).

9 See G. Rattray Taylor, *op. cit.*; S. J. Gould, *op. cit.*

10 See for instance Jacques Monod, *op. cit.* Monod was designated a Nobel Prize-winner for his work in molecular biology. See further, Appendix 10.

11 I am not claiming that it could be demonstrated by scientific procedures, but only stressing how much is involved.

12 *op. cit.,* p. 110. Jacques Monod would have claimed to be an atheist. His book has been a 'phenomenal best-seller', and has been 'the subject of major comment throughout the world'.

13 *Ibid.,* p. 113 (his italics).

14 *Ibid.,* p. 114.

15 *Ibid.,* p. 131.

16 Monod would have been the last to claim that his remarks were intended only for 'symmetrical' situations. There are relatively few such.

17 For instance, (in the Kinetic Theory of Gases), by measuring the distribution of molecular velocities and showing that it follows Maxwell's equation.

18 The case of the telephone numbers is very much to the point here.

19 *op. cit.,* p. 111.

20 This, of course, would make the experiment meaningless!

21 Science knows nothing of any other modes of distinguishing.

22 For one thing, there are too many established physical (i.e. scientific) phenomena for which we have hardly a clue, such as the migration of freshwater eels back to their distant breeding grounds at the end of their lives.

23 i.e. by so-called psychokinesis.

24 That is, that there is no relationship with the direction of evolution.

25 Or other genetic phenomena such as recombination.

26 His reaction would be rather like that of the blockhead who complained, 'I don't know what I want but I shan't be satisfied till I get it.'

27 Unless he slips it in under cover of the Uncertainty Principle; for a criticism of this suggestion, see W. G. Pollard, *Chance and Providence* (Faber and Faber, London, 1959).

28 John 4:48; Matt. 6:26,30.

29 Job 26:14 (RSV, NIV); Psalm 29.

30 The opposite claim, it should be noted, is not fundamental to the biblical position.
31 The numbers are *statistically* random. This is a concept wider and more demanding than that which the biologists require!

NOTES TO CHAPTER 12 (pages 97 to 106)

1 For instance, the failure of the 'bound feet' of Chinese women to influence subsequent generations.
2 For simplicity in what follows we neglect the paired nature of genes. The omission is immaterial.
3 In our model, if height depends on only two 'marbles' there will be only three grades (both colourless, one of each, both black) to cover the whole range from dwarf to tall.
4 For a popular fair-minded review see Alan Hayward, *Creation and Evolution* (Triangle, 1985). See also Fred Hoyle and C. Wickramasinghe, *Evolution from Space* (Dent 1981).
5 D. J. Batholomew, *op. cit.,* Chapter 5.
6 I myself find it incredible.
7 Secker and Warburg 1983. The quotation is from p. 236. It is an example of what the Greeks called *hubris,* a spirit often rebuked in the Bible: cf Job. 38:1ff, Isaiah 45:9ff, 1 Cor. 1:20ff. *Evolution: a theory in crisis* by Michael Denton (Burnett Books, 1985) is an excellent and even more recent book. The author is a molecular biologist who writes professionally and without religious reference.
8 *The Material Basis of Evolution* (Yale University Press 1940). Goldschmidt's ideas were generally ridiculed.
9 See further, Appendix 11.
10 Rattray Taylor, *op. cit.,* p. 86 puts the figure nearer 100 to one.
11 The same twenty or so amino-acids form the common constituents of proteins from all organisms; they are all in the same (l-) stereoisomeric form (out of the two possible); adenosine triphosphate is a universal energy metabolite; chlorophyll is related to the haeme of blood; and so on. Not surprisingly, cell ultrastructure shows many universal features.

NOTES TO CHAPTER 13 (pages 107 to 114)

1 Heb. 3:7; 10:15ff.
2 John Calvin, *Commentary on Psalms,* Ps. 136:7; quoted by R. Hooykaas, in *Christian Faith and the Freedom of Science* (Tyndale Press, 1957).
3 Exod. 20:8–11; Deut. 5:12–15; Mark 2:27.
4 Eccles. 3:14c.
5 'Evening came and morning came'—cyclically (Gen. 1 NEB) cf. Eccles. 1:1–10; 3:15; Matt. 1:2ff. Ecclesiastes draws some important lessons from this cyclicity (3:20; 11:9,10; 12:14).
6 Pss. 90; 102:23–27 (compare the force of 'days' and 'years' in vv. 23,24).
7 D. Kidner, *op. cit.,* pp. 48ff on Gen. 1:11,20,24.
8 Cf. the NEB rendering of Gen. 1:21 ('every kind of bird') and the GNB

rendering of vv. 11,12,20,21,24. The text *cannot* be construed as stating positively the creation simultaneously or *sui generis* of every species.

9 Gen. 2:7,19.

10 cf. C. K. Barrett, *The Gospel according to St. John* (SPCK, 1978), p. 570. The Creed refers to the Holy Spirit as the 'Lord and giver of life'; cf. Job 33:4; John 6:63; Rom. 8:2,11.

11 For the last, see Luke 1:80; Acts 7:23, 34.

12 Ezek. 36:22f.

13 Rom. 5:12.

14 For the meaning of the primal sin see Appendix 5.

15 1 Cor. 15:55,56.

16 Language can be learned only from others who already know it.

17 cf. H. Blocher, *op. cit.,* p. 96 '. . . it is our encounter with another which allows our inner life to become aware of itself'.

18 Letter to J. S. Henslow, 11 April 1833.

19 Letter to Charles Whitley, 23 July 1834.

20 *Life and letters of Charles Darwin* Vol. III ed. Francis Darwin (John Murray, 1887).

21 I believe I owe this illustration to Prof. D. M. MacKay.

22 All of these characeristics can be traced in the lower animals—see W. H. Thorpe *Animal Nature and Human Nature* (Methuen, 1974); J. Z. Young, *Introduction to the Study of Man* (Oxford U.P., 1971).

23 Isaiah 28:23ff. cf. Gen. 2:15.

24 Rom. 7:9,10 (NIV).

25 cf. John 9:41; 2 Pet. 2:21; Rom. 3:20. John 15:21b–25 bears strongly on the same theme. It explains the strong antipathy ('hatred') commonly felt by men and women for God *as Lawgiver* (note the frequent reference to 'commandments' in this chapter), arising from ignorance of him (v. 21b), and quite 'without cause'.

26 Heb. 9:27; cf. 1 Cor. 15:56.

27 Rom. 5:19,18,12,14 (RSV and NIV).

28 Deut. 29:29; Rom. 15:4–6.

29 See Chapter 11.

30 As, for instance, by the orthogeneticists.

31 Job 11:7; 23:8,9; Ps. 77:19; Isaiah 45:15; Rom. 11:33; 1 Tim. 6:16 cf. Luke 10:21.

32 Bernard Lovell, *In the Centre of Immensities* (Paladin Books, 1980), pp. 122, 156.

NOTES TO CHAPTER 14 (pages 115 to 121)

1 Ronald W. Clark, *The Survival of Charles Darwin* (Weidenfeld and Nicolson, London, 1985).

2 *op. cit.,* p. 5.

3 Autobiography, p. 114.

4 Quoted in a letter written (1884) after his death by Frances Julia Wedgewood to his own son Francis Darwin—see R. W. Clark, *op. cit.* p. 195.

5 For instance, the widespread acceptance of the notion of the fixity of species probably owed a lot more to the influence of Plato and Aristotle than to the Bible—see Michael Denton (*op. cit.*).

6 Matt. 6:26ff.

7 cf. G. von Rad, *op. cit.*, pp. 26, 98: 'This is anything but the bluntness and naïveté of an archaic narrator. It is, rather, the candour and lack of hesitation which is only the mark of a lofty and mature way of thinking.' 'Its simplicity, however, is not archaic, but rather the highest command of every artistic means.'

8 We need to be clear here. The Bible does validate the picture of the potter and the clay (Is. 6:48; Jer. 18:6); but it uses it (in a different connection) as a *metaphor,* not as a model. A metaphor remains limited in its scope; a model is designed to relate what we are talking about to the totality of knowledge bearing on the same subject. That is why the model (unlike the metaphor) must 'grow' as knowledge grows. It is worth remarking that the same verb (*yāṣar,* form) is used of the divine activity in Amos 7:1, where God is 'forming' locusts (apparently out of grass!) by the normal process of biological growth.

9 The reader will find many suggestive thoughts in Dorothy Sayers, *The Mind of the Maker* (Methuen, 1942).

10 As would a full-scale working model of a locomotive!

11 If the idea of mechanism or method belongs (biblically speaking) anywhere, it belongs to Providence. Evolution may be the method of Providence, but not of Creation. See further, Appendix 3(i).

12 Isaiah 65:17,18; John 3:5; Rev. 22:1–5.

NOTES TO EPILOGUE (pages 122 to 124)

1 Methuen, 1974 (p. 370).

2 Thorpe quotes this expression from Tillich.

3 Matt. 11:27; John 1:18. (We are speaking of 'knowing' in the sense defined in Job 42:5 by 'sees').

4 Matt. 11:25.

5 1 Cor. 1:21, 2:10.

6 Gen. 1:1; Ps. 36:9.

7 John 1:1,3,9,14.

8 Col. 1:17.

NOTES TO APPENDIX 1 (pages 127 to 131)

1 I am indebted for this argument to Paul Helm (*Themelios* 4, 20–24, 1978).

2 B. W. Anderson ed., *Creation in the Old Testament,* (SPCK, 1984).

3 No value judgement is implied by the application of the epithet 'lower' to questions about what may be the correct text of a document and 'higher' to discussion of its authorship, composition, date and purpose.

4 'inspired by God' (RSV); 'God-breathed' (NIV); 2 Tim. 3:16.

5 Rom. 9:17; Dan. 5:18–23; John 19:10,11. cf. also 2 Sam. 7:8; Isaiah 44:28–45:6 and *passim*.

6 Further, within this general perspective one such government may have a unique relationship with God, different from that of all others. The Christian mind thinks of Jesus Christ—'My King, upon my holy hill of Zion' (Ps. 2:6).

(This note is added lest the reader should think that the argument implies that all literature is in this connection equal.)
7 BCP *Thirty-Nine Articles,* Art. 20.

NOTES TO APPENDIX 2 (pages 131 to 135)

1 In this discussion 'twenty-four hours' stands for the length of a twentieth century day.
2 So E. H. Andrews, *God, Science and Evolution* (Evangelical Press, 1980), p. 126.
3 Gen. 1:14–18. Compare Ps. 139:11,12.
4 Ps. 74:16 (note the reference to Creation). In a very common biblical usage the joining of 'night' and 'day' vividly conveys this idea of *continuousness* (e.g. Gen. 7:4; Lev. 8:35; Deut. 28:66,67; Ps. 77:2; Acts 9:24; 20:31; Rev. 4:8). That is no doubt its purpose here.
5 Matt. 5:48; John 5:19; 17:4; Rom. 12:11; Eph. 5:1.
6 Rom. 12:1 (AV). The Greek adjective is *logikos.* NEB and RSV translate, [the worship] 'offered by mind and heart', 'spiritual' [worship].
7 John 17:4; 19:30; Col. 1:20–22 (both 'reconciles' are aorists, signifying completed action).
8 2 Cor. 5:20.
9 'Aeons' or 'ages'.
10 cf. its use in Matt. 4:21; 21:16; 1 Cor. 1:10; Heb. 10:5.
11 e.g. Matt. 11:25; Luke 16:31; 24:25; John 12:36,37; 2 Cor. 4:4.
12 Heb. 11:4; Gen. 4:2f.
13 The related passage in Heb. 1:2 lends its support to all this. 'Created the world' (RSV) is here literally 'made the aeons' (*epoiēsen tous aiōnas*), with the verb *poieō* (make, do) corresponding to the Hebrew *'āśâ* of Gen. 1:7,16,25,26,31; 2:2,3,4. It makes good sense here too to interpret the 'aeons' as the Genesis days.
14 See H. Blocher, *op. cit.,* p. 57.
15 Psalm 104:26,31 cf. v. 34.
16 'Man's chief end is to glorify God and enjoy Him for ever' (The Shorter Catechism). See Isaiah 58:13ff; Heb. 4:10,11.
17 *'Āśâ* (cf. French *faire*) is a common verb meaning both 'do' and 'make'.
18 Followed by its articulation as *fiat.*

NOTES TO APPENDIX 3 (pages 135 to 137)

1 It is well expressed in the Collect for Trinity 8 in the Book of Common Prayer: 'O God, whose never-failing providence ordereth all things both in heaven and earth; . . .'
2 In 2 Pet. 3:5 *sunistēmi* is represented by the perfect participle; in Col. 1:17 by the perfect indicative. In this verb the perfect is present in meaning. The verb is rarely used intransitively in the NT, the only other case being Luke 9:31.
3 NIV. The literal rendering of the verbs here is, 'are having-been-reserved being-kept'.
4 This is the form following the singular feminine noun 'creature' (Gen. 1:24 RV). Masculine and plural forms are similar.

NOTES TO APPENDIX 4 (pages 137 to 138)

1 D. Kidner, *op. cit.,* p. 49; U. Cassuto, *op. cit.,* pp. 49, 50.
2 Where *tannînim* (or its singular) is translated 'serpents', 'dragons', 'monster' in RSV. Note also the association with 'fierce' meteorological phenomena in Psalm 148:7,8.

NOTES TO APPENDIX 5 (pages 139 to 142)

1 Gen. 3:5,22.
2 *Op. cit.,* p. 132 cf. 2 Sam. 14:17; 1 Kings 3:9.
3 J. Calvin, *Commentary on Genesis* (1554), p. 118.
4 G. von Rad. *op. cit.,* p. 89.
5 *Ibid.,* p. 97.
6 Matt. 26:38,39; John 6:38; 8:29; Heb. 10:5,7.
7 Deut. 32:20 (RV); Hab. 2:4; John 3:16; Rom. 1:17.
8 Deut. 6:4,5; Ps. 91·9,14; John 14.15; Gal. 5:6; 1 Tim. 1:5; 1 John 4:16; 5:3.
9 Henri Blocher, *op. cit.,* p. 184.
10 Alan Hayward, *op. cit.,* p. 199 (my italics).
11 Deut. 30:19,20.
12 Cf. our Lord's words in Matt. 10:28.
13 Job 18:14.
14 1 Cor. 15:56; cf. also Heb. 9:27.
15 See also John 5:24; 11:26; Rom. 8:2,6.
16 In fact, Scripture records that it did (approximately). This may be a clue to the significance of this otherwise enigmatic period; see the references in Rev. 20:1–10. Isaiah foretells that in the messianic 'new heavens and new earth' the earthly life of the redeemed will be of this order (Isaiah 65:20,22). On this supposition the righteous will live the millennium through.
17 1 Cor. 15:50–54; Phil. 3:20,21.
18 Heb. 6:20.
19 Cf. Luke 24:52,53.
20 As our Forerunner, restoring what Adam lost, our Lord never 'saw' this—Acts 2:24,27.

NOTES TO APPENDIX 6 (pages 143 to 144)

1 See the comments of D. Kidner, *op. cit.* p. 82.
2 G. von Rad, *op. cit.,* p. 72.
3 'Genealogy', *New Bible Dictionary* (IVP, 1980).
4 G. Wenham, *Numbers, An introduction and Commentary* (IVP, 1981).
5 See further the comments and note at the end of Appendix 5.

NOTES TO APPENDIX 7 (pages 144 to 146)

1 See Matt. 6:33; 10:28; Luke 12:15; John 6:27; 2 Cor. 4:18.

2 This point is often made forcefully in Scripture: cf. Gen. 19:24; Isaiah 10:22,23 quoted in Rom. 9:27,28; Nah. 1:7–9. See also 2 Pet. 2:5–10.

3 Gen. 7:19,20.

4 2 Pet. 3:6 RV; Isaiah 28:17–22; Jer. 11:11.

5 Heb. *har* means a mountain or hill. In Gen. 22:2 quite a small hill is clearly indicated, probably the site of the future temple. Cf. also 'mount Zion'. The common word *'erets* is twice as often translated 'land' as 'earth'.

6 Alan Hayward, *Creation and Evolution* (Triangle Books SPCK, 1985); Dan Wonderly, *God's Time Records in Ancient Sediments* (Crystal Press, Michigan 1977); Davis Young, *Christianity and the Age of the Earth* (Zondervan, Michigan 1982).

7 Num. 14:26–32; Isaiah 10:21–23; Matt. 24:37f.

8 For a fuller discussion see the article 'Flood' in *The New Bible Dictionary* (IVP, 1980); also *Genesis* D. Kidner (IVP, 1967).

NOTES TO APPENDIX 8 (pages 146 to 147)

1 This may be illustrated by the analogy of a human operator who uses a biased die. He has designed the bias to ensure that 90% of the throws are sixes. But the results of individual throws he remains unable to predict.

2 In *A Working Faith*, 1980. See also D. J. Bartholomew, *op. cit.*

3 This conclusion can perhaps be evaded: see the interesting discussion in Batholomew, *op. cit.,* pp. 81,82.

4 It would be very difficult to establish, in a long series of coin tosses, that now and then the coin had been caught in mid-air and deliberately placed.

5 D. M. MacKay, *The Clockwork Image* (IVP, 1974); *Science, Chance and Providence* (OUP, 1978).

6 D. J. Bartholomew, *op. cit.,* pp. 23–25.

NOTES TO APPENDIX 9 (pages 148 to 149)

1 As Job, Jeremiah and Habakkuk did—and were answered with hope and encouragement; see Job 38:1f; Jer. 12:1,14f; Hab. 1:2f; 2:2.

2 Matt. 5:11,12; Rom. 5:3–5; 2 Cor. 12:9,10; Heb. 12:11; cf. John 16:33.

3 Rom. 8:20,21.

4 Rom. 8:18; 1 Cor. 15:54.

5 The 'new world', Matt. 19:28.

6 Ps. 16:11.

7 See Eph. 3:17–19; Rev. 21:3,4.

8 John 10:17,18; 12:26; 15:9–11; 16:20–22.

9 Rom. 5:8; John 15:13.

10 Luke 15:11f.

11 Is this the significance of Paul's amazing and profound statement in Rom. 11:32?

12 John 3:16; Rom. 8:32.

13 Rom. 8:30; Heb. 2:10.

14 Rom. 8:35–39; cf. also Rev. 1:5,6.

15 Rev. 5:6; 22:1. The 'Lamb . . . slain' is a symbol of Jesus Christ crucified and raised from the dead. Cf. also Luke 7:42,43,47.

16 Eph. 3:18; John 17:11,25.

17 Matt. 25:34; 1 Cor. 2:2,7,8; Eph. 1:4–10.

NOTES TO APPENDIX 10 (pages 150 to 153)

1 Second edition (1963), p. 54.
2 *Ibid.* p. 576.
3 Revised edition (1967), p. 164.
4 *Ibid.* pp. 292, 293, 345.
5 Scientific American *239* (1978) pp. 46–55.
6 *The Thread of Life* (G. Bell, 1966).

NOTES TO APPENDIX 11 (page 153)

1 *Seven Clues to the Origin of Life* (CUP, 1985).

NOTES TO APPENDIX 12 (pages 154 to 155)

1 See, for instance, Deut. 30:19; Matt. 12:36.
2 i.e. radial distance from himself.
3 The idea of dimensions here is, of course, much wider than usual. It includes not only space and time coordinates and such physical dimensions as mass, temperature and electrical potential, but others relating to personal and supra-personal elements of reality. A 'dimension' may accordingly be defined as any respect for which an 'entry' would have to be made in order completely to specify an occasion. (See Daniel Lamont, *Christ and the World of Thought*, T. and T. Clark, Edinburgh, 1935; also the discussion in W. H. Austin, *The Relevance of Natural Science to Theology*, Macmillan, 1976.)
4 For further discussion see D. M. MacKay, *op. cit.* and W. G. Pollard, *op. cit.*

For Further Reading

The literature on both the biblical and scientific sides is voluminous. On the former the following are recommended:

Blocher, Henri *In the Beginning* (IVP, 1984). A profound, luminous and scholarly study by a French professor, of the meaning of the opening chapters of Genesis (the sections likely to be difficult to a layman are in smaller print).

MacKay D. M. *Science, Chance and Providence* (Oxford University Press, 1978). A brilliant defence by an eminently lucid, original and logically-careful writer of the orthodox, biblical, Christian view of God's activity in nature and history.

—*The Clockwork Image* (Inter-Varsity Press, 1974). A book addressed to both Christians and non-Christians concerned to know what biblical faith would mean for their intellectual integrity in an age of mechanistic science.

There are a number of able books very critical of evolutionary orthodoxy; the following can be recommended:

Taylor, Gordon Rattray *The Great Evolution Mystery* (Secker and Warburg, 1983). An able and provocative review of the difficulties facing orthodox Darwinism by a writer who is apparently an atheist; well illustrated.

Hoyle, Fred and Wickramasinghe, Chandra *Evolution from Space* (J. M. Dent, 1981). A fascinating account by the eminent cosmologist (an erstwhile atheist) and a mathematician of their view of the inadequacies of Darwinian theory, and of their own highly unorthodox theory of how life both originated and continued to evolve under cosmic influences presided over by intelligence.

Denton, Michael *Evolution: a Theory in Crisis* (Burnett Books, 1985). A very professional, well-informed and readable account by a molecular biologist (apparently without religious prejudices) of the difficulties of orthodox theory. Well illustrated and with some striking and original analogies.

Some works of more general interest are:

Russell, Colin A. *Cross-currents: Interactions between science and faith* (Inter-Varsity Press, Leicester, 1985). An excellent and very readable and authoritative account of the inter-relations of Christian Faith and Science, including Darwinism.

Lovell, Bernard *In the Centre of Immensities* (Granada, 1980).

A work of 'brilliant analysis and passionate humanity' by a great radio-astronomer, this deals in ultimate terms with Man's place in the Cosmos.

Midgley, Mary *Evolution as a Religion* (Methuen, 1985). A professional philosopher (not anti-Darwinian nor obviously pro-Christian) discusses the extravagant mythologies which threaten to discredit not only Darwinian theory but science at large. She probes very deeply into modern secularist faiths.

Durant, John (ed.) *Darwinism and Divinity* (Blackwell, 1985). A collection of interesting and perceptive essays written from a variety of standpoints.

Index of Biblical References

184   _Index of Biblical References_

Index of Names and Subjects